Noise and Representation Systems:
A Comparison among Audio Restoration Algorithms

This volume is humbly and affectionately dedicated to the memory of Gian Antonio Mian (1942-2006) – Full Professor of Digital Signal Processing with the Department of Information Engineering, Padova University (Italy) – a leading researcher and an outstanding teacher whose brightness and kindness I will always remember

The book is listed in http://www.lulu.com, where it is free downloadable and where reviews can be posted.

Printed by lulu.com.

Cover design: Sergio Canazza
A deterministic noise (hum) representation by means of Lissajous' figure (curves obtained by the superposition of different simple harmonic oscillations in orthogonal directions). These figures are similar to the pictures made by a Spirograph toy, but they are made by measuring audio or radio signals.

In the late 19th century by the French physicist Jules Lissajous invented a method of graphical investigation of the superposition of two perpendicular vibrations. He used sounds at different frequencies to vibrate a mirror. A beam of light reflected from the mirror would trace patterns which depended on the frequencies of the sounds (a setup similar to the apparatus which is used today to project laser light shows in multimedia environment).

Before the days of digital frequency meters and phase-locked loops, Lissajous' figures were used to determine the frequencies of noise signals. A signal of known frequency was applied to the horizontal axis of an oscilloscope, and the noise signal to be measured was applied to the vertical axis. The resulting pattern was a function of the ratio of the two frequencies. The device used – since the mid-19th century – for this purpose was the harmonograph (Turner, 1997), a mechanical apparatus that employs pendulums to create the Lissajous' patterns. A simple harmonograph uses two pendulums to control the movement of a pen relative to a drawing surface. One pendulum moves the pen back and forth along one axis and the other pendulum moves the drawing surface back and forth along a perpendicular axis. By varying the frequency of the pendulums relative to one another (and phase) different patterns are created. More complex harmonographs incorporate three or more pendulums or linked pendulums together (for example hanging one pendulum off another).

Contents

But surpassing all stupendous inventions, what sublimity of mind was his who dreamed of finding means to communicate his deepest thoughts to any other person, though distant by mighty intervals of place and time! Of talking with those who are in India; of speaking to those who are not yet born and will not be born for a thousand or ten thousand years.
[Galileo Galilei, *Dialogue Concerning the Two Chief World Systems*, 1632]

People who make no noise are dangerous.
[Jean de La Fontaine, *Fables*, Book VII, fable 23, 1678-1679]

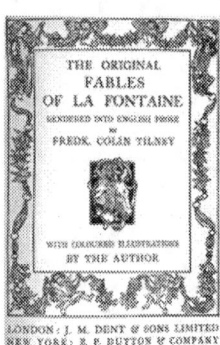

Preface

Over the last ten years research in the audio restoration field has focused on the developing of algorithms, which subtend a plurality of models and hypothesis on the sound reality and have been designed in connection with the peculiar problem which the system intends to solve:

- restoration can aim at retrieving the intelligibility of the spoken parts during a communication occurring in a perturbed environment, where the critical factor is real-time working, even if at the expense of a high loss of vocal timbre quality (communication between pilots and control tower, between divers and the mother ship, or between troops on enemy soil);
- a commercial objective can, instead, concern the understanding of the spoken parts during a communication in a noisy environment. In this case, real-time working of the system is essential, but with an at least partial preservation of the original timbre (communication between *mobile* devices in shopping centres, or at parties, concerts and the like);
- intelligibility retrieval is also the aim of restoration in the forensic field. In this case, real-time working is not required, but faithfulness to timbre must be guaranteed for the identification of the vocal print;
- this essay wants to analyse in detail those algorithms specifically dedicated to the restoration of musical recordings, which must offer satisfying solutions to the problems connected with the time-varying feature peculiar to musical signals. Real time is not required (the work of the restorer often requires 10 or 20 times real time). In this case the purpose of the restoration is determined according to the various document orientations adopted (Orcalli 2006), therefore noise reduction interventions could: 1) concern only the cases in which the internal evidence of the degradation (*documental* approach) is unquestionable, without going beyond the technological level of that time; 2) aim at a commercial edition (*aesthetical* approach); 3) have the purpose of obtaining a historical reconstruction of the recording as it was listened to at the time (*sociological* approach);

9

4) have the objective to preserve the *intention* of the author (*recostructive* approach).

The methodologies specific to audio restoration can be schematized in at least three different categories, according to the information used by the algorithm during the noise attenuation phase.

1) *Frequency methods;* these algorithms require that the operator have little information to carry out the restoration (a *priori* information): only a noise estimation is necessary (*noise print*), since it is assumed to be stationary along the entire signal. Any information further information needed (a *posteriori* information) is calculated automatically by the restoration software through the analysis of the characteristics of the signal. Since these algorithms are easy to use and are generally applied to different typologies of audio signals, they are employed in commercial hardware and software systems.

2) Algorithms in the time domain, which use *signal models*; employ *a priori* information to estimate the probability distribution of sound events, the excitation signal and filter coefficients. Therefore the algorithm carries out (a *posteriori* information) the signal tracking. The models, which can be applied to different signal typologies are *non-informative* (they have little *a priori* information): it is therefore necessary to detail the model according to the signal being examined.

3) Restoration through *analysis by synthesis* and restoration based on *source models*; in these cases only *a priori* information is required. It is to be found in the knowledge we have about the system that produced the audio document and the analysis of the sound material.

The signals produced by optoelectronic devices are affected by statistical uncertainty due to discrete nature both the support and the electromagnetic radiation. In this sense, the noise can be defined as a spontaneous process able to mask or interfere with the intentional signal.

The degradation that an audio recording can be subject to consists of a series of disturbances, which can be briefly classified in two groups:

1) deterministic noise: *hum* (typically caused by ground loops that inject a 50Hz or 60Hz signal into the *intentional signal* cables), *HVAC* (noise generated by electrical Heating, Ventilation and Air Conditioning), high frequency *harmonic noise* (caused by: a- imperfections in the

wiring of power supply; b- some noise source in the electrical power system like light dimmer; c-magnetic fields from transformers of not audio equipment);

2)   stochastic noise, grouped in:

- local disturbances;
- global disturbances.

Local disturbances affect only a portion of the musical signal and can be identified as a discontinuity in its waveform. The deriving effects can vary according to the physical condition of the recording carrier and depend on factors, such as the signal frequency, and the duration and localisation of the disturbance. This category encompasses, for example, the numerous *ticks*, which can be perceived when listening to common vinyl records. They are characterised by a short duration and are principally caused by the granularity of the material (vinyl, for instance) and by the presence of dust in on the microgrooves. Then there are the superficial scratches and other disturbances (pops, breakages, and clippings) characterised by longer temporal extensions. Similar artefacts affect also other analogue carriers such as films, wax cylinders, and magnetic tapes. It is important to highlight that certain ticks, due to jitters or writing errors, can be noticed even in digital recordings.

The global disturbances, instead, affect the whole signal and include broadband background noise, wows, flutters and some kinds of non-linear distortions. Background noise is perhaps the most common type of degradation and is perceived by the listener as a hiss. This disturbance is caused by the electric and mechanical parts of the equipment used for the recording, by the environment and the physical deterioration of the carrier. In practice it occurs whenever the signal is present in analogue form. Its filtering is an integral part of the work carried out in the recording and post-production studios and is usually carried out before the digital conversion takes place.

From the point of view of restoration, the main problem is represented by the fact that such noise has significant components at every frequency of the audio spectrum. The risk of significantly altering the useful part of the signal is not at all remote. Thus, adequate algorithms should be studied.

In Section I, the various approaches used for noise removal will be presented. In particular the classic methodologies for the individuation and removal of local and global disturbances will be described (1.2 e 1.3). In Section II, some innovative algorithms, developed by the author in cooperation with Gian Antonio Mian of the Padova University, will be illustrated in detail. First of all, the algorithms based in the frequency domain will be explained (2.1), then those based in the time domain, which use signal models (2.2).

It is very difficult to identify evaluation criteria for the results of the various audio restoration algorithms. Section III will focus on some numeric indexes, which aim at indicating the quality of a musical restoration (3.1) and will show the results obtained during the restoration of a piece by using the above-described algorithms: in (3.2) by using the algorithms in the frequency domain and in (3.3) by using the algorithms in the time domain. In Section IV an example of restoration through *analysis by synthesis* is given.

An Italian version of this text is available in book:
S. Canazza and M. Casadei Turroni Monti (eds.). (2006). *Ri-mediazione dei documenti sonori*. Udine: Forum.

Acknowledgement

I would like to thank Angelo Orcalli (Full Professor with the Department of Scienze Storiche e Documentarie, Udine University, Italy) for the meticulous and tireless discussion on this topics; Giovanni De Poli (Full Professor with the Department of Information Engineering, Padova University, Italy) for a large number of excellent suggestions; Engineer Luca Novello for the experimental work carried out; Valentina Clama for the English translation, and Deborah Saidero (Assistant Professor with the Department of Lingue e Letterature Germaniche e Romanze, Udine University, Italy) for the linguistic supervision.

This work has been carried out with the support of the CULTURE 2000 Programme of the European Union.

1. Standard algorithms

1.1 Model-based Approach

The first restoration techniques used in the past were analogue and required the manual editing of magnetic tapes through a *cut and paste* procedure in order to remove the local defects (i.e. scratches) and a frequency equalization to reduce global degradations, such as hisses and harmonic distortions. Since impulsive disturbances have a high content of high frequency harmonics better results were soon obtained by carrying out a high-pass filtering, in order to detect their presence and, secondly, by proceeding with the removal through low-pass filters. None of these methods is however sufficiently sophisticated to allow for a significant degree of reduction of the disturbances without somehow compromising the quality of the original signal.

The recent development of digital techniques, boasting a higher degree of precision and a higher operational flexibility, allows for sensibly better results. However such techniques, if used improperly, can also cause a sensible worsening of the audio document. New investigations in the field of restoration through numeric process are based on the deployment of mathematic-statistical models to describe the sound source – source models (Esquef, Välimäki, and Karjalainen, 2002) – or the audio signal and the different types of disturbances under consideration – signal models (Bari, Canazza, De Poli, and Mian, 2001). More specifically, the audio has to be regenerated according to the model employed, the choice of which represents, in this case, a critical point. In the electro-acoustic music field, for example, sinusoidal wave combinations with frequency modulation, with central frequency and time-varying amplitude are often used. In this case, it is possible to conceptually hypothesize to plan an appropriate generative model of such signals with frequency modulation. The exact number of components, their frequencies and amplitudes could be generated following an opportunely chosen probability distribution (according to the compositional model used in that given piece), in order to synthesize artificial signals ideally identical to those used by the composer and/or performer. These random generated quantities represent the *incognito* parameters of the model, which are to be determined during the analysis phase of the existing signal.

How the quality of the restoration largely depends on the capacity of the model chosen to faithfully describe the audio signal is inferable. Suppose to denote with *x(t)* the value of the musical signal sampled at the time *tT*, where *T* represents the time between two subsequent digital samples. It is therefore possible to assume the existence of a process capable of generating these digital samples which can be written in the form *x(t)=f(θ, e(t))*, where the vector *θ* represents the model parameters, *e*, is a random noise, or *excitation* term which describes any casual element present in the signal, e *f(.)* is a function which enables us to pass from the knowledge of the parameters and the excitation to the value of the audio signal samples.

The term concerning the excitation *e(t)* can be interpreted as a random deviation of the real signal from the idealised model, even before being considered as the effect of a random environmental noise or a noise in some way extraneous to the original signal. The vector of *θ* parameters could be time-varying, in order to take into consideration temporal changes in the features of the signal considered.

In literature the most used model is the autoregressive (AR) model with constant coefficients of *p* order, (poles model)

$$x(t) = \sum_{i=1}^{p} x(t-i)a_i + e(t) \tag{1}$$

The current value of the discrete- time signal *x(t)*, *t*=1, 2, ... is expressed through a linear combination of its previous *p* samples to which the excitation term *e(t)* is added. This model is able to aptly represent both harmonic signals (poles close to the unitary circle) and noisy signals (poles close to the origin), and the *p* order can be interpreted as a complexity index of the waveform which has to be studied. Since the characteristics of the audio signal satisfy the stationary hypothesis only locally, the necessity to update the a_i coefficient at regular time intervals (~20ms) is to be taken into account, that is to say, a time-varying coefficient model (TVAR) has to be adopted.

A model (McAulay and Quartieri, 1986) similar to that related to the previous example is the sinusoidal one, which exploits an additive synthesis of sinusoids with frequency modulation to represent the signal:

$$x(n) = \sum_{i=1}^{p_n} a_1(n) \sin\left(\int_0^{nT} \omega_i(t)\,dt + \varphi_i \right)$$

In this way is possible to obtain a frequency and amplitude modulation, since $a_i(n)$ and $\omega_i(t)$ vary in time. Furthermore it is possible to describe the *birth* and the *death* of single frequency components, since p_n also varies in time. The biggest disadvantage of this solution consists in the extreme complexity of the parametric identification phase.

In addition, this model is not very apt to represent signals having characteristics similar to noise.

1.2   Local disturbances

Depending on whether the click is added to the musical signal (additive model) or replaces it completely for a short time period (substitutive model), the signal affected by local disturbances can be differently modelled. Usually the additional model is considered acceptable for commonly found defects (which can be commonly found), such as ticks caused by dust and small scratches, whereas the substitutive model is necessary to describe more serious deteriorations of the recording carrier. Henceforth, reference will be made to the following model of general validity proposed in Godsill (1993; 1995):

$$y(t) = x(t) + d(t)v(t) \tag{2}$$

where $y(t)$ represents the available signal, $x(t)$ the non-noisy signal, $v(t)$ is the single click and $d(t)$ an indicating function which has unitary value in correspondence to the samples affected by disturbance and zero elsewhere. The statistic of the $d(t)$ process decides when a sample is damaged, while that of $v(t)$ determines the typical disturbance amplitude. Obviously the value of $v(t)$ has no influence when $d(t)$ is null and remains irrelevant even if $d(t)=1$, in case of a complete substitution of the corrupted samples.

Although different approaches can be adopted to handle impulsive noises, in order to obtain the maximum quality all methods should ideally intervene only on the damaged samples leaving the others unaltered. This aspect leads to the identification of two different phases in the restoration process:

- The detection phase, in which the position and duration of the disturbances present in the useful signal are identified;
- The removal phase, in which the deteriorated audio signal is reconstructed as accurately as possible.

In reality, on the basis of psycho-acoustic considerations, it will be sufficient to remove only the artefacts perceivable by the human ear, having the purpose to reach the best compromise between disturbance removal and the inevitable distortion which occurs carrying out the processing.

1.2.1   Detection

As previously mentioned, one of the simplest methods to detect clicks consists in carrying out a low-pass filtering on the signal. Usually, in fact, in contrast with impulsive disturbances, common audio material has little information for frequencies exceeding 8 kHz. The detection phase could therefore be carried out applying a waterfall threshold detector to the low-pass filter. This method, which has been employed both in analogue systems (Carrey and Buckner, 1976; Kinzie and Gravereaux, 1973) and digital systems (Kasparis and Lane, 1993), has the advantage to be easily implemented and does not require the estimate of peculiar parameters, apart from the threshold value. Some problems could arise when the disturbances have a limited band or when elaborating musical signals with a rich high frequency content.

A similar approach can be used exploiting the wavelet theory and by using a technique which is substantially equivalent to filtering by means of a bank of filters (Montresor, Valiere, Allard, Baudry, 1990; Montresor, Valiere, Allard, and Baudry, 1991; Valiere, 1991). However, for our purposes more interesting are the methodologies, which include some a priori information on the quantities considered, by representing the audio signal through an AR model (Vaseghi and Rayner, 1988; Vaseghi and Rayner, 1990; Vaseghi, 1988) and by exploiting the principles of robust filtering (Efron and Jeen, 1992). In order to understand the idea at the basis of these techniques, consider that the trend of the musical signal *s(t)* can be effectively described through a locally stationary AR process having an input as white noise *e(t)*. Hypothesising that we have obtained from the noisy signal *y(t)* a sufficiently reliable estimate of the parameters a_i of the model, applying for this purpose any robust parametric identification procedure to impulsive disturbances, from the equations (1) and (2) we can obtain the following expression for *e(t)*:

$$e(t) = y(t) - \sum_{i=1}^{p} x(t-i)a_i - d(t)v(t) \tag{3}$$

In absence of disturbances (*d(t)=0*) results, of course, *y(t)=s(t)* and (3) can be simplified in:

$$e(t) = y(t) - \sum_{i=1}^{p} x(t-i)a_i \tag{4}$$

which underlines how applying to y(t) the *whitening* filter:

$$A(z) = 1 - \sum_{i=1}^{p} a_i z^{-i}$$

we can easily obtain the trend of e(t), which in this case must have the characteristics of white noise.

On the basis of this observation, it is possible to believe, at least theoretically, that the clicks can be identified by carrying out a simple local *whitening* test on the output signal *e(t)* from the filter (4), by employing a threshold detector and supposing that every irregularity in the noise trend can indicate the presence of a disturbance.

This scheme makes it possible to amplify the relation between disturbances and non- noisy signals, but at the expense of precision in the temporal click localisation, which now involves p+1 e(t) samples owing to the intrinsic structure of the whitening filter (4). Such an imprecision can became critical in presence of very close disturbances with very different amplitudes, since the effects can be reciprocally added or eliminated. The choice of the threshold value, which is influenced by the amplitude of the clicks, has to be the outcome of a compromise, which enables for the individuation of the highest number of disturbances without incurring too often into false alarms.

There are also methods which employ matched filters adapted to the impulsive disturbances and, in this case, considering the audio signal as additive coloured noise, try to detect their presence (Van Trees, 1968). Ultimately, there are other methods exploiting the theory of neural networks (Czyzewski, 1997).

1.2.2   Removal

The classical impulsive disturbance removal strategy is based on the complete reconstruction of the damaged samples through interpolation techniques. Since it is often impossible to retrieve any useful information from the degraded fragment, its integral replacement constitutes the simplest choice and coincides with that adopted herein. There are however more sophisticated procedures (Godsill, 1993; Godsill and Rayner, 1995), which try to retrieve information even from the portion of the signal affected by the disturbance through noise modelling methods.

The problem can be formalised in the following way: consider N samples of the audio signal which constitute a x vector. y is the correspondent vector containing the noisy signal and d is that indicating presence of the clicks. The x vector can be divided into two parts: the first, which contains the elements whose value is known (that is in which $d(t)=0$) will be denoted with x_k; while the second, concerning the corrupted samples and therefore unknown ($d(t)=1$) will be denoted with x_u. Similarly the vectors y and d are partitioned. All this can be seen as the problem of x_u estimate starting from the y data observed.

Different methods were developed for this purpose; the simplest is that of average filter (Ptas and Venetsanopoulos, 1990; Kasparis and Lane, 1993; Nieminen, Miettinen, Heinonen and Nuevo, 1987). In practice, the damaged elements are replaced with an average value which takes into account the characteristics of the signal waveform. This approach is however not apt to handle fragments longer than 10 samples.

More effective methods are surely those based on the definition of a model, which exploit the information deduced a priori from the characteristics of the signal in order to assess its trend starting from the available measures. In particular in the following pages the autoregressive interpolator will be mentioned (Ptas and Venetsanopoulos, 1990; Kasparis and Lane, 1993; Nieminen, Miettinen, Heinonen, and Nuevo, 1987). This technique, called Least Squares AR-based (LSAR), was developed by Vaseghi and Rayner and was originally employed for the removal of digital ticks in CDs and DATs.

Consider a block of N samples which presumably lacks the x signal generated as output by a (locally stationary) AR process having a parameters. Suppose to have retrieved an assessment of the AR model coefficients starting from the available samples shortly before and after the damaged fragment. Expressing $e(t)$ of the (1) equation in a matrix form we have:

$$e = \boldsymbol{A}x$$

in which $\boldsymbol{A}$ is a matrix having the dimensions $(N\text{-}p)\times N$ in which the line $(j\text{-}p)$ is so constructed to generate the residue :

$$e(j) = x(j) - \sum_{i=1}^{p} x(j-i)a_{j,i} \tag{5}$$

The second element of this equation can be divided, as previously explained, into sections including the known and unknown data, so that $\boldsymbol{A}$ results partitioned into columns. The least square solution is obtained minimising the index:

$$E = e^T e \tag{6}$$

As we know in this case the solution can be expressed in a closed mathematical formula. Various algorithms which can numerically solve the problem are available. The experimental data show how the LSAR approach always produces results of a higher quality than the pure interpolators (which do not formulate assumptions on the nature of the signal), and the choice of high orders for the AR model can contribute (within certain limits) to better reconstruct the wider portions of the deteriorated signal.

This interpolator has the property of being the minimum variance non-polarised estimator of the failing samples (Veldhuis, 1990). Observe that such a procedure requires however the presence of a robust parametric identification procedure for the impulsive disturbances (Huber, 1981), and represents therefore a sub-optimal choice.

The problem becomes sensibly more complex if we suppose that even the *a* vector of the AR model coefficients is incognito. In this case, the minimisation of the equation (6) compared with x_u and *a*, corresponding to the maximum likelihood estimator of the missing samples and of the parameters, contains undetermined factors of the fourth order and cannot be analytically resolved. A possible solution (Jassem *et al.*, 1986) could consist in operating before the data and after the parameters: this would guarantee a convergence, at least locally, of the maximum likelihood function.

This approach, which will be described in detail hereunder, was initially put forward by M. Niedzwiecki and K. Cisowski (Niedzwiecki, 1989; Niedzwiecki and Cisowski, 1996) to solve the problem by simultaneously exploiting the *Extended Kalman Filter* (EKF).

However we have to highlight the fact that, whatever the techniques employed, the use of *overmixing* procedures can highly improve the quality of the restored sound (Czyzewski, 1997). They consist in carrying out two different assessments of the damaged samples: the first elaborates the data *foreword*, in the portion of the signal preceding the click; the second is instead carried out inverting the temporal axis (proceeding *backwards*) in the portion which follows the disturbance. The final restoration will be then calculated mediating the two reconstructions on the basis of a certain optimum criterion (Etter, 1996) or by using , for example, a Kalman filter as *smoother* (Lewis, 1986).

To conclude the study of impulsive noise, the peculiar typology known as *low frequency impulsive* noise need also be mentioned. Although it is normally grouped with other disturbances, its substantial difference must be considered. Its duration is, in fact, superior to that of normal clicks (it can reach some 0.1 second) and it is caused by a marked defect, even by the break of the recording carrier. A typical case in records is the heavy discontinuity, which induces an oscillation in the head with a high content of low frequency energy. Another peculiarity is the fact that there is *overlapping* with the audio signal and not a *replacement* of it. The algorithms planned for the clicks cannot eliminate these defects in a satisfying way. The innovative solution presented by Godsill (1993) includes a reduction based on *template*. The advantage is given by the fact that the shape of the disturbance is almost identical on the entire surface of the carrier; the characteristics of a *representative* taken in a portion, where the musical signal is absent, are memorised. It will thus suffice to carry out a sort of temporal subtraction where the disturbance added to the useful signal occurs. This method is not easy to implement automatically (the phase of disturbance detection is very difficult owing to the high presence of low frequency energy) and therefore it is mainly used in a manual way.

1.3   Global disturbances

The additive broadband noise is the most common form of global disturbance and, having significant spectral components at every frequency, cannot be eliminated through simple equalisation procedures. In practice its stationariness can often be hypostasised and, in order to avoid overvaluation problems, it is better to carry out the filtering only after having removed the possible impulsive disturbances from the musical signal.

Over the last decades many techniques concerning background noise reduction in audio recordings have been developed. Many of these share the fact of being based on a certain type of a priori knowledge related to the signal and/or the noise.

A very schematic classification, which can be traced among the existing restoration methods, consists in the division between the algorithms that act in the frequency domain and those which act in the time domain. The first and most widespread methods are the frequency domain methods. Usually these do not refer to a model associated to the signal and can thus be implemented and used more easily. On the contrary, the methods based on the time domain often refer to a signal model and therefore require a higher experience for their use, but they enable us to potentially obtain better results in the quality of restoration.

1.3.1   Frequency domain methods

The most widespread techniques employ a signal analysis through the Short-Time Fourier Transform (which is calculated on small partially overlapped portions of the signal: Short-Time Fourier Transform STFT) and can be considered as a non-stationary adaptation of the Wiener filter in the frequency domain. In particular, short-time spectral attenuation (STSA) consists in applying the short-time spectrum of the noise to a time-varying suppression and does not require the definition of a model for the audio signal.

Suppose to consider the useful signal $x(t)$ as a stationary aleatory process to which some noise $z(t)$ is added (uncorrelated with $x(t)$) to produce the degraded signal $y(t)$:

$$y(t) = x(t) + z(t)$$

The relation that connects the respective power spectral densities is therefore:

$$P_y(\omega) = P_x(\omega) + P_z(\omega)$$

with w the frequency index.

If we hypothesize to succeed in retrieving an adequate estimate of $P_z(\omega)$, during the silence intervals of the signal $y(t)$, and in the musical portions that of $P_y(\omega)$, we can expect to obtain an estimate of the $x(y)$ spectrum by subtracting $P_z(\omega)$ from $P_y(\omega)$ (Boll, 1979; Boll, 1991; Lim and Oppenheim, 1978; Lim and Oppenheim, 1979); the initial assumption of stationariness (assumption) can be considered locally satisfied since short temporal windows are employed.

Note that the use of a short-time signal analysis is equivalent to the use of a filter bank. First each channel (that is, the output of each filter) is appropriately attenuated and then it is possible to proceed with the synthesis of the restored signal. The time-varying attenuation applied to each channel is calculated through a determined suppression rule, which has the purpose to produce an estimate (for each channel) of the noise power. Each particular STSA technique is characterised by the way in which the filter bank is carried out and the suppression rule defined.

Often the short-time analysis is carried out through the STFT (Lim and Oppenheim, 1979; Ephraim and Malah, 1984; Cappé, 1991). In MacAulay and Malpass (1980), instead, non-linear filter banks are introduced.

Historically, the STSA methodology was developed during the '70s in order to remove the noise in the transmission of spoken parts. The new STSA techniques for audio restoration are an adaptation of these first processes. Traditionally, the interpretation as STFT is a notion deriving from the analysis of the spoken parts.

This phase remains an open problem: in the STFT interpretation, the attenuation corresponds to a change of the short-time spectrum module only. The opinion that the phase does not need to be processed owing to the psycho-acoustic properties of the human ear is widespread. Indeed, the *insensitivity to the phase* of the human ear is proved only in the case of stationary audio signals and for the Fourier Transform phase. On the contrary, in the case of the STFT phase, variations among subsequent short-time frames can cause audible effects (such as frequency modulation).

It is important to highlight that in the classic STSA techniques the possibility to process the phase does not exist, since no hypothesis is made on the characteristics of the audio signal.

If we denote the STFT of the *y(t)* noisy signal with $Y(t, \omega_k)$, where *t* represents the temporal index and ω_k the frequency index (with K=1,...,N: N represents the number of STFT channels), the result of the suppressing rule application can be interpreted as the application of a $G(t, \omega_k)$ gain to each value $Y(t, \omega_k)$ of the STFT of the noisy signal. This gain corresponds to a signal attenuation and has to be included between 0 and 1.

In most of the suppression rules, $G(t, \omega_k)$ depends only on the noisy signal power level (measured in the same point) $|Y(t, \omega_k)|^2$ and on the estimate of the noisy power at the ω_k frequency,

$$\hat{P}_z(\omega_k) = E\left\{ \left| Z(t, \omega_k) \right|^2 \right\}$$

(which does not depend on the temporal index *t* due to the presumed noise stationariness).

At this point a *relative* signal can be defined

$$Q(t, \omega_k) = \frac{|Y(t, \omega_k)|^2}{\hat{P}_z(\omega_k)} \tag{7}$$

which, starting from the hypothesis that the *z(t)* noise is not correlated to the *x(t)* signal, we deduce to be always superior to 1:

$$E\{Q(t, \omega_k)\} = 1 + \frac{E\left\{ \left| X(t, \omega_k) \right|^2 \right\}}{\hat{P}z(\omega_k)}$$

A typical suppression rule is based on the Wiener filter (Wiener, 1949) and can be formulated as follows:

$$G(t, \omega_k) = \frac{|Y(t, \omega_k)|^2 - \hat{P}_z(\omega_k)}{|Y(t, \omega_k)|^2}$$

Other rules, such as the power-subtraction, are illustrated in Boll (1979), Oppenheim (1979).

In Figure 1 the characteristics of two rules are compared in connection with the relative signal $Q(t, \omega_k)$. From Figure 1 we deduce that the suppression rules share the same behaviour:

- $G(t, \omega_k) = 1$, where the relative signal is high ($Q(t, \omega_k) \gg 1$)

- $\lim_{Q(t, \omega_k) \to 1} G(t, \omega_k) = 0$. That is, the gain tends to 0 in the case in which only the noise is present (relative signal equal to 1). In this sense, in some cases an overvaluation of the estimated noise power is used.

Other more elaborated suppression rules depend on both the relative signal and on a priori knowledge of the corrupted signal, that is to say, on a priori knowledge of the probability distribution of the under-band signals (Boll, 1191) or on the signal to noise ratio (Ephraim and Malah, 1984). Usually the mistake made by these procedures in retrieving the original sound spectrum has an audible effect, since the difference between the spectral densities can give a negative result at some frequencies. Should we decide to arbitrarily force the negative results to zero, in the final signal there will be a disturbance, constituted of numerous random frequency pseudo-sinusoids, which start and finish in a rapid succession, generating what in literature is known as *musical noise* (Ephraim and Malah, 1984).

In the following pages such suppression rules will be explained in detail together with some changes and innovations to them which will be presented by using also psychoacoustic considerations concerning the masking phenomenon.

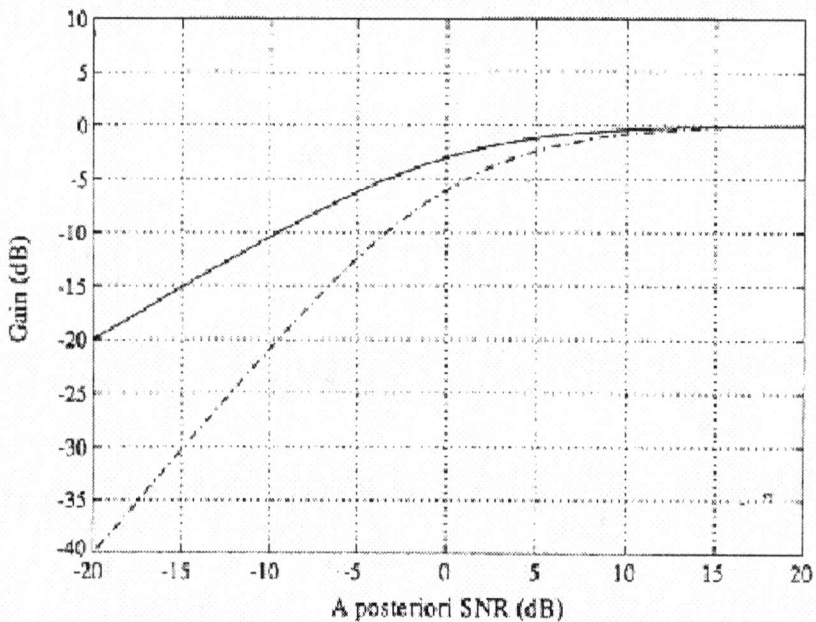

Fig. 1: *Comparison between Wiener filter and power-subtraction*

1.3.2 Time domain methods

Although global disturbances are traditionally treated through the frequency method, as those above-mentioned, recently methods based on a time domain approach and referring to an AR model to represent the musical signal have been developed.

A first method in the time domain, which will be analysed in detail in the following sections, proposes to solve the three principal aspects of restoration (parametric identification, click detection/removal and background noise attenuation) in a unified manner, by treating them as an integrated non-linear filtering problem for whose solution a series of already consolidated tools can be used, among which the Kalman filter (KF). This general scheme solves, in a recursive way which is particularly apt for a numeric implementation, the problem of the estimation of a linear model state. Based on equations characterized by a two-phase mechanism (prediction-updating), the different ways in which it can be employed make it possible to carry out operations, such as signal filtering, the parameter tracking of a model, or the interpolation of missing samples.

The Kalman filter boasts some important properties: first of all, it is the linear estimator, which minimises the estimation error variance. If the probability distributions considered are Gaussian, it also becomes the minimum variance estimator in an absolute sense. Its application to the treatment of non-linear models (called Extended Kalman Filter or EKF) also carries out, during each step, an *intelligent* linearization of the model around the best estimation of the state available at that moment, thus making it possible to limit the errors due to such operations. The main consequence deriving from the employment of this approximation consists in the loss of optimality and convergence guarantees.

A second method, which will be illustrated, equally operating in the time domain and based on an AR model of the signal, uses the equations of the Kalman filter, but only during the background noise attenuation phase. The aspect concerning the identification of the parameters associated with the model employed is, in fact, treated by using statistic tools, such as the Monte Carlo method. This approach consists in generating a high number of realisations for the value of the parameters associated with the model and in a subsequent selection of them on the basis of the probability they have of being correct. Once an estimation of the parameters is obtained, the signal model associated with them is called into play and the Kalman filter equations are applied so as to obtain an estimation of the original signal.

2. Innovative algorithms

2.1 Frequency domain methods

2.1.1 Ephraim-Malah filter (EMSR)

After the Wiener solution, many variants, which are also affected by musical noise, even if in a minor way, were proposed. On the contrary a substantial progress was made with the solution hereinafter proposed. The work, carried out in Ephraim and Malah 1984, aims at minimising the mean square error (MSE) in the estimation of the spectral components (Fourier coefficients) of the musical signal, of which A_k indicates the module:

$$E\left\{\left(A_k - \hat{A}_k\right)^2\right\}$$

By modelling Ak as a statistically independent null mean Gaussian aleatory variables, the obtained solution is:

$$\hat{A}_k = \Gamma(1.5)\frac{\sqrt{v_k}}{\gamma_k}\exp\left(-\frac{v_k}{2}\right)\left[(1+v_k)I_0\left(\frac{v_k}{2}\right)+v_k I_1\left(\frac{v_k}{2}\right)\right]R_k \qquad (8)$$

where:

$$v_k = \frac{\xi_k}{1+\xi_k}\gamma_k$$

$$\gamma_k = \frac{Y_k^2}{E\left[|Z_k|^2\right]} \qquad (a\ posteriori\ \text{SNR}),$$

$$\xi_k = \frac{E\left[|X_k|^2\right]}{E\left[|Z_k|^2\right]} \qquad (a\ priori\ \text{SNR}),$$

and X_k, Z_k, Y_k are the spectral components of the clean signal $x(t)$, of the noise $z(t)$ and of the noisy signal $y(t)$ respectively. I_0 e I_1 are the Bessel modified functions of zero and one order. Note that the quantity ξ_k can only be estimated, since the clean signal is not available. The estimate calculation is developed according to two models, one based on a maximum likelihood approach and the other based on a decision directed approach. Since the latter one turned out to be the best, we report it here (n is the frame index):

$$\hat{\xi}_k(n) = \alpha \frac{\hat{X}_k(n-1)}{Z_k(n-1)} + (1-\alpha) P\big[y_k(n) - 1\big] \qquad\qquad 0 \le \alpha < 1 \qquad\qquad (9)$$

$$P[x] = \begin{cases} x & \text{if } x \ge 0 \\ 0 & \text{otherwise} \end{cases}$$

In Cappé, 1994 the behaviour of the filter based on such an estimator is analysed; after a notation change the gain applied to each spectral component k to the p-nth frame is:

$$G(k,p) = \frac{\sqrt{\pi}}{2} \sqrt{\left(\frac{1}{1+Y_{post}(k,p)}\right)\left(\frac{Y_{prio}(k,p)}{1+Y_{prio}(k,p)}\right)} \cdot M\left[\left(1+Y_{post}(k,p)\right)\left(\frac{Y_{prio}(k,p)}{1+Y_{prio}(k,p)}\right)\right]$$

$$M[\vartheta] = \exp\left(-\frac{\vartheta}{2}\right)\left[(1+\vartheta)I_0\left(\frac{\vartheta}{2}\right) + \vartheta I_1\left(\frac{\vartheta}{2}\right)\right]$$

where the two parameters Y_{post} e Y_{prio} are calculated as:

$$Y_{post}(k,p) = \frac{|X(k,p)|^2}{v(k)} - 1$$

$$Y_{prio}(k,p) = (1-\alpha)P\big[Y_{post}(k,p)\big] + \alpha \frac{|G(k,p-1)X(k,p-1)|^2}{v(k)}$$

$$P[x] = \begin{cases} x & \text{se } x \ge 0 \\ 0 & \text{otherwise} \end{cases}$$

where $v(k)$ is the noise power at the k frequency. The α parameter controls the balance between the current frame information and that of the preceding one. By varying this parameter, the filter smoothing effect can be regulated. In Figure 2(a) it is possible to observe that Y_{prio} (that is the SNR calculated taking into account the information of the preceding frame, continuous line) has less variance than Y_{post} (broken line). In this way, it is less probable that a musical noise occurs.

Furthermore, considering Figure 2(b) it is possible to observe the effect of the variation of α from 0.98 to 0.998; we note an estimate of Y_{prio} about 10dB lower and a decisively minor variance. This results in a more effective musical noise reduction. However, it is important to point out that when α increases, the response delay is even higher than the transistors; therefore there is a low-pass effect on the occasion of rapid signal attacks.

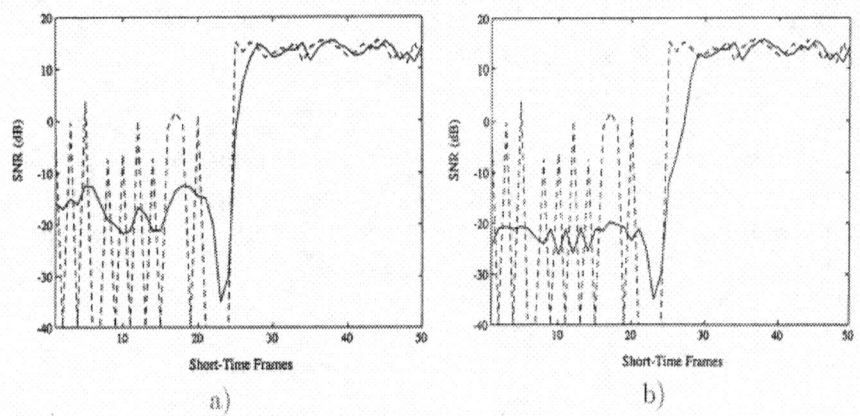

Fig. 2: *Influence of the a parameter (from 0.98 to 0.998).*

A good adaptation to the non-stationary noise case (see Cappé, (1994) par. III B) is another advantage of the proposed algorithm; this is particularly important in the case of the restoration of vinyl records or Shellac discs (78 rpm), where the noise can be modelled as cycle-stationary with an excellent approximation, with a period equal to that of the record rotation (Godsill and Rayner, 1998).

We have also to consider that increasing the overlapping of the analysis windows the statistic correlation degree between the frames increases. This results in a limitation to the noise reduction power of the filter. The analysis carried out in Canazza, De Poli, Maesano, and Mian

(1999) concluded that a hard overlapping (exceeding 80%) can give acceptable results only by increasing the value of *a*.

2.1.2   *Ephraim-Malah with logarithmic estimator*

In literature an evolution of the EMSR suppression rule, based on an estimator which minimizes the logarithmic mean square error in the estimation of the signal spectral components was presented, that is:

$$E\left\{\left(\log A_k - \log \hat{A}_k\right)^2\right\}$$

where A_k represents the module of the *k*-nth Fourier coefficient. The estimator obtained in Ephraim and Malah (1995) is:

$$\hat{A}_k = \frac{\xi_k}{1+\xi_k}\exp\left\{\frac{1}{2}\int_{v_k}^{\infty}\frac{e^{-t}}{t}\,dt\right\}Y_k$$

In the above-quoted article, this realisation is presented as a significant evolution of the standard EMSR: in particular it produces a definitely inferior musical noise at the expense of a minor uniformity (however it is hardly audible due to the minor remaining noise).

2.1.3   *Innovative filters based on EMSR*

This paragraph presents the implementation of some filters created by the authors which represent an evolution of the Ephraim and Malah suppression rule.

The first one, EMSR *<alpha>*, is based on the idea of using a *punctual* suppression without memory, Wiener typology, in the case of a null estimate of Y_{post}; the pseudo-code is the following:

IF Y_{post}(k,p)>0
  $\alpha = 0.98$

ELSE
  $\alpha = 0$

END

The experiments carried out in Canazza, *et al.* (1999), confirm that the filter performs very well during the listening phase, with a noise removal decidedly better than the classic EMSR (about 5dB reduction) and the prerogative of not introducing musical noise, at least for *SNR* exceeding 20dB. Furthermore the behaviour in the transistors follows that of EMSR without having the impression of a "low-pass filter" application.

The second algorithm, called *EMSR <alpha-past>*, is the evolution of the previous one and takes into account the information of the last two frames. More precisely, before brutally setting the parameter at zero, we observe if the preceding frame also contained a null Y_{post}. The pseudo-code is:

IF $Y_{post}(k,p)>0$

$\qquad \alpha = 0.98$;

$$Y_{prio}(k,p) = (1-\alpha)P[Y_{post}(k,p)] + \alpha \frac{|G(k,p-1)Y(k,p-1)|^2}{v(k)} ;$$

ELSE

$\qquad$ IF $Y_{post}(k,p-1)>0$

$\qquad\qquad \alpha = 0.98$;

$$Y_{prio}(k,p) = (1-\alpha)P[Y_{post}(k,p-1)] + \alpha \frac{|G(k,p-2)Y(k,p-2)|^2}{v(k)} ;$$

$\qquad\qquad Y_{post}(k,p) = Y_{post}(k,p-1);$

$\qquad$ ELSE

$\qquad\qquad \alpha = 0$

$\qquad$ END

END

Observing the behaviour during the listening phase, has pointed out the minor noise attenuation in comparison with the EMSR *<alpha>*. In addition, the analysis also showed a substantial reaction delay in the descent times (passage from signal only to noise only).

The last algorithm, called *EMSR <past>*, to calculate Y_{prio} uses the following formula:

$$Y_{prio}(k, p) = 0.98 * \left[(1-\alpha) P[Y_{post}(k, p)] + \alpha \frac{|G(k, p-1) X(k, p-1)|^2}{v(k)} \right]$$

$$+ 0.02 * \left[(1-\alpha) P[Y_{post}(k, p-1)] + \alpha \frac{|G(k, p-2) X(k, p-2)|^2}{v(k)} \right]$$

As we can observe, in order to estimate Y_{prio}, the decisional strategy of using the past previous frames *p*-1 and *p*-2 at 2%, that is only as corrective parameters, was adopted. The behaviour during the listening phase emphasized a minor noise reduction in comparison with the two preceding algorithms.

2.1.4 Perceptive filter

The perceptive filter here presented was implemented in Canazza *et al.* (1999), Tsoukalas, Mourjopoulos, and Kokkinakis (1997) and in Beerends and Stemerdink (1992). The general scheme is represented in Figure 3 .

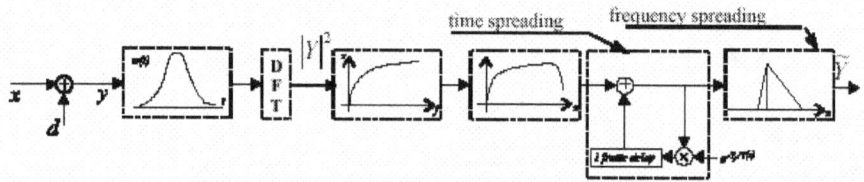

Fig. 3: *Scheme of the psychoacoustic model*

The following operations are represented in order:
- windowing in time and DFT with the relative calculation of the noisy signal power (*x(t)* is the clean signal and *d(t)* is the noise);
- passage from the Hertz scale to the Bark scale through the estimation of the relative signal excitation at each critical band;
- outer to inner ear transformation;

- time spreading which is an operation with memory of the preceding frame;
- frequency spreading.

The $\tilde{Y}$ signal so obtained is the psychoacoustic representation of the *y(t)* signal; the same transformations are carried out on the available sound print. With these representations of the noisy signal and of the noise, a filtering is carried out with a suppression rule similar to that of Wiener. This solution shows scarce aptitude to generate musical noise, in comparison with the Wiener filter in the *external* domain, thanks to the spreading operations which contribute to attenuate the filtering variance, that is, to render its trend smoother and less irregular. For this purpose see Figure 4 where the gains of psychoacoustic and of the Wiener filters are compared (in the external domain).

In order to employ psychoacoustic criteria in noise removal, the use of a suppression rule, which considers the masking effect which the noise was subjected to, is necessary. The spreading thresholds which present the original signal *x(n)* are not known *a priori* and are to be calculated. This estimation can be obtained by applying a noise reduction STSA standard technique leading to an estimate in the frequency domain of the original signal $\hat{X}$ *(k)*, for which the masking thresholds m_k, defined as the non negative threshold under which the listener does not perceive an additional noise, can be calculated by using an appropriate psychoacoustic model.

The masking effect obtained is incorporated into one of the STSA standard techniques taking into consideration the masking thresholds m_k for each *k* frequency of the STFT transform. A cost function depending on m_k, which minimised gives the suppression rule for the noise reduction, is created. This cost function can be a particularisation of the mean square deviation to include the masking thresholds, under which the cost of an error is equal to zero.

Defining with $\hat{a}_k$ the estimation of the spectral amplitude a_k of the original signal , the cost function used is:

$$C\left(\hat{a}_k, a_k\right) = \begin{cases} \left(\hat{a}_k - a_k\right)^2 - m_k^2 & \text{if} \quad \left|\hat{a}_k - a_k\right| > m_k \\ 0 & \text{otherwise} \end{cases}$$

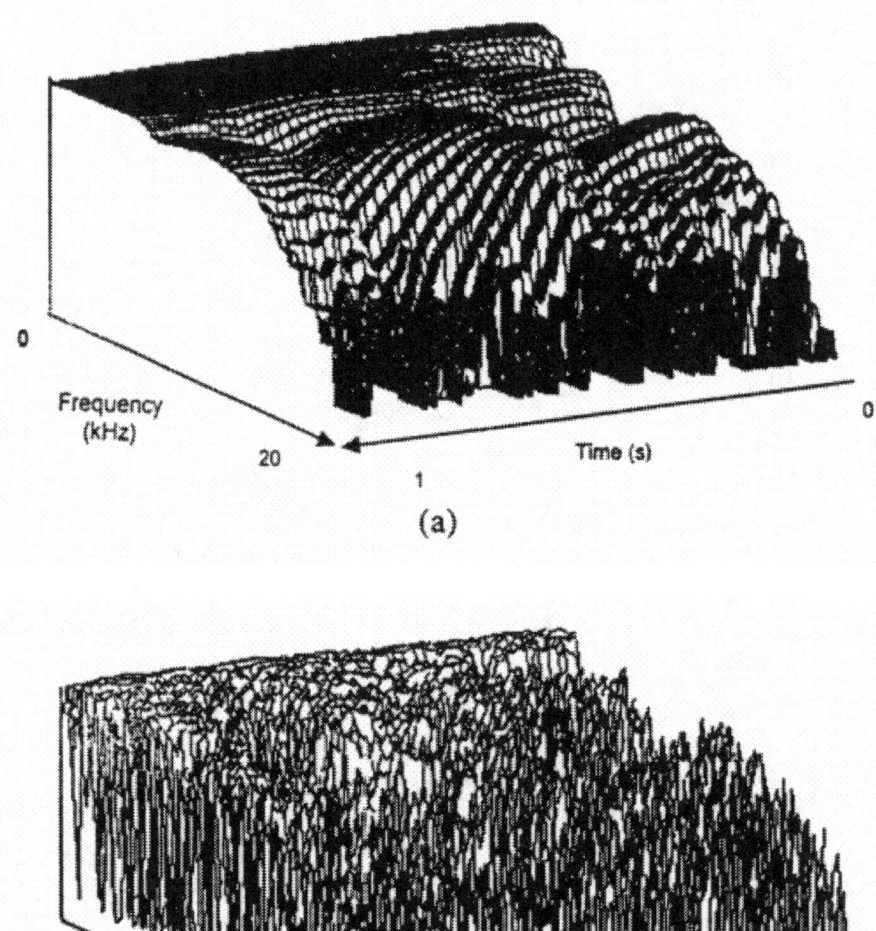

Fig. 4: *Comparison beween the gains of the psyhcoacoustic and pseudo Wiener filters*

The cost function, given the observed value Y(k) and the masking threshold m_k, is minimised calculating

$$\min_{\hat{a}_k} \int_{a_k} \int_{\alpha_k} C(\hat{a}_k, a_k) p(a_k, \alpha_k | Y(k)) da_k da_k \qquad (10)$$

The mathematical resolution, carried out with numeric methods, leads to a suppression rule shown in Figure 5 as a function of the instantaneous signal to noise ratio SNR $\xi(k)$ for different masking levels. We can immediately notice how such a suppression law tends asymptotically to the masking threshold m_k for low signal to noise ratio levels. Furthermore, in the case of $m_k = 0$, the amplitude estimator is once again the traditional original filter (from which we started), in the case examined the Ephraim and Malah filter.

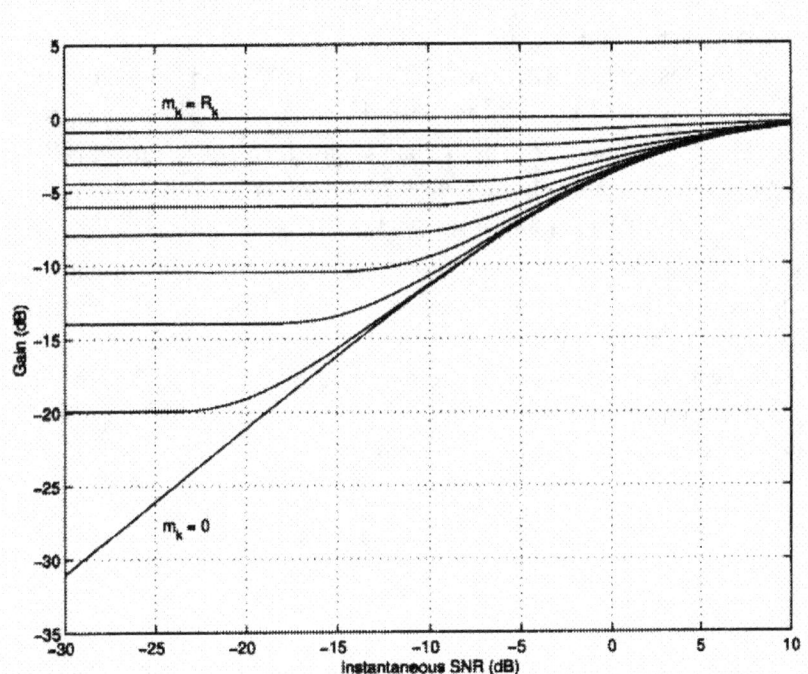

Fig. 5: *Suppression rule in function of the istantaneous signal to noise ratio SNR for different masking levels obtained analytically.*

The application of the suppression rule which employs the perceptive model applied to the Ephraim and Malah filter requires numerous and computationally heavy mathematical operations. In order to obtain a discrete computational efficiency it is preferable to use an approximation of the optimal solution. The Ephraim and Malah suppression rule, calculated as the function of the *a priori* signal to noise ratio SNR $\xi(k)$, with an inferior limit which avoids attenuation under the masking level m_k is used: if the spectral amplitude Y_k of the observed signal is minor to the masking threshold, the signal is not modified, since the spectrum is masked by the original signal. Such approximation can be expressed as:

$$H(k) = \begin{cases} \max\left[\dfrac{\hat{m}_k}{R_k}, \dfrac{\hat{\xi}(k)}{1+\hat{\xi}(k)} \right] & \text{if} \quad R_k > \hat{m}_k \\ 1 & \text{otherwise} \end{cases}$$

whose graphic is shown in Figure 6.

Even the estimation of the original signal, used to calculate the masking levels, is carried out by using the Ephraim and Malah suppression law, intended as A function of the *a priori* $\hat{\xi}$ signal to noise ratio SNR , but calculated using a parameter higher than α compared with that used for the calculation of $\hat{\xi}$ in (9). This variation enables us to obtain an estimation of the original signal with much less residual noise, but with an inevitably much higher signal attenuation. This is not desirable if the objective is signal restoration, but it is decidedly hoped for in the calculation process of the masking levels associated with the estimate of the spectral components of the original signal, because it allows for an estimation of the threshold taking into account only the original signal.

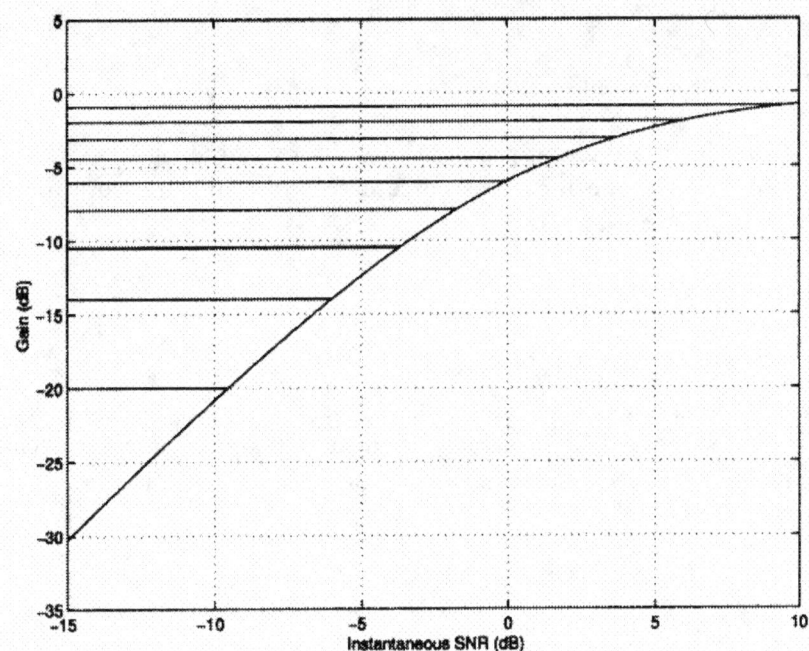

Fig. 6: *Approximated suppression law in function of the instantaneous signal to noise ratio SNR for different masking levels.*

2.2 Time-domain methods

2.2.1 Extended Kalman Filter

Consider the filtering problem of a discrete-time signal $s(t)$, $t=1,2,\ldots$ having at disposal the measures corrupted by a white measure noise $z(t)$ (the background noise).

$$y(t) = s(t) + z(t) \tag{11}$$

The original audio signal $s(t)$ is modelled as an AR process of p order with known time-varying coefficients $a_1(t),\ldots,a_p(t)$ according to model (1).

Then assume that the white processes $z(t)$ and $e(t)$ are Gaussian and mutually uncorrelated (therefore statistically independent), that is:

$$z(t) \sim N(0,\sigma_z^2)$$
$$e(t) \sim N(0,\sigma_e^2)$$
$$z(t) \perp e(t)$$

Considering these hypothesis, the problem of the $s(t)$ retrieval from the noise measures $y(t)$ can be solved optimally through the Kalman filter. Introducing the notation:

$$\theta(t) = \begin{bmatrix} a_1(t) \\ \vdots \\ a_p(t) \end{bmatrix} \quad \varphi(t) = \begin{bmatrix} s(t) \\ \vdots \\ s(t-p+1) \end{bmatrix}$$

and

$$A(t) = A[\theta(t)] = \begin{bmatrix} a_1(t) & \ldots & a_{p-1}(t) & a_p(t) \\ 1 & \ldots & 0 & 0 \\ \vdots & \ddots & \vdots & \vdots \\ 0 & \ldots & 1 & 0 \end{bmatrix} \quad c(t) = \begin{bmatrix} 1 \\ 0 \\ \vdots \\ 0 \end{bmatrix}$$

the equations (1) and (11) can be rewritten in the form:

$$\begin{cases} \varphi(t+1) = A(t)\varphi(t) + ce(t) \\ y(t) = c^T \varphi(t) + z(t) \end{cases} \tag{12}$$

which represents the standard formulation in the state space of the signal filtering problem.

The optimal prediction in a stage and the optimal estimate of the $\varphi(t)$ vector starting from the past measures $Y(t) = \{y(t),...,y(1)\}$ are given by:

$$\hat{\varphi}(t+1\,|\,t) = E\big[\varphi(t+1)\,|\,Y(t)\big] \qquad\qquad \textit{Optimal prediction}$$
$$\hat{\varphi}(t\,|\,t) = E\big[\varphi(t)\,|\,Y(t)\big] \qquad\qquad\quad \textit{Optimal Estimate}$$

and can be calculated recursively using the Kalman filter whose equations, in this case, become:

- **Prediction phase:**

$$\begin{cases} \hat{\varphi}(t+1\,|\,t) = A(t)\hat{\varphi}(t) \\ \Sigma_\varphi(t+1\,|\,t) = A(t)\,\Sigma_\varphi(t\,|\,t)\,A^T(t) + cc^T \end{cases} \tag{13}$$

- **Updating phase:**

$$\begin{cases} \hat{\varphi}(t\,|\,t) = \hat{\varphi}(t\,|\,t-1) + L_\varphi(t)\,\varepsilon_\varphi(t) \\ \Sigma_\varphi(t\,|\,t) = \left(I_p - L_\varphi(t)\,c^T\right)\Sigma_\varphi(t\,|\,t-1) \\ \varepsilon_\varphi(t) = y(t) - c^T\,\hat{\varphi}(t\,|\,t-1) \\ L_\varphi(t) = \dfrac{\Sigma_\varphi(t\,|\,t-1)\,c}{c^T\,\Sigma_\varphi(t\,|\,t-1)\,c + k_0} \end{cases} \tag{14}$$

where $\varepsilon_\varphi(t)$ is the prediction error, $L_\varphi(t)$ the Kalman gain and $k_0 = \dfrac{\sigma_z^{\,2}}{\sigma_e^{\,2}}$ is

the scalar noise to signal ratio.

Since:

$$p(\varphi(t+1) \mid Y(t)) = (\hat{\varphi}(t+1 \mid t), \sigma_e^2 \Sigma_\varphi(t+1 \mid t))$$

$$p(\varphi(t) \mid Y(t)) = (\hat{\varphi}(t \mid t), \sigma_e^2 \Sigma_\varphi(t \mid t))$$

$\Sigma_\varphi(t+1 \mid t)$ and $\Sigma_\varphi(t \mid t)$ can be interpreted as covariance matrices of the prediction and filtering error respectively, *normalised* in comparison with the variance of the input noise σ_e^2.

   With some small changes to the previous equations even the problem concerning the elimination of impulsive noises can be solved. Suppose that an external variable $d(t)$ is available and acts as an indicator of the presence of an impulsive noise:

$$d(t) = \begin{cases} 0 & \text{absence of impulsive noise} \\ 1 & \text{absence of impulsive noise} \end{cases} \tag{15}$$

The following time-varying relation is introduced in replacement of the constant ratio k_0:

$$k(t) = \begin{cases} k_0 & d(t) = 0 \\ \infty & d(t) \neq 0 \end{cases}$$

Setting $k(t) = \infty$ we indicate that the corresponding measure is corrupted by an impulsive noise and therefore has no useful information for the restoration of the original signal.

The values assumed by the Kalman gain and the estimators in equation (14), in this case become:

$$\begin{cases} L_\varphi(t) = 0 \\ \hat{\varphi}(t \mid t) = \hat{\varphi}(t \mid t-1) \\ \Sigma_\varphi(t \mid t) = \Sigma_\varphi(t \mid t-1) \end{cases}$$

with the effect that the measure $y(t)$, although available, is ignored by the filter.

2.2.1.1   *Smoothing*

The optimal estimate, in the sense of the least square, of $s(\tau)$ given the Y(t) measures assumes the known form:

$$\hat{s}(\tau|t) = E[s(\tau)|Y(t)]$$

If $t > \tau$, the quantity $\hat{s}(\tau|t)$ is called the smoothed estimate of $s(\tau)$ and at the τ time combines not only all the *past* measures but also a certain number of *future* measures (obviously introducing a decision delay of t-τ samples).

The Kalman filter given by the equations (13) and (14) acts as a smoother with a fixed observation interval (equal to p). In fact we have:

$$\hat{\varphi}(t|t) = \begin{bmatrix} E[s(t)|Y(t)] \\ \vdots \\ E[s(t-p+1)|Y(t)] \end{bmatrix} = \begin{bmatrix} \hat{s}(t|t) \\ \vdots \\ \hat{s}(t-p+1|t) \end{bmatrix}$$

and the last variable included in the state vector at the t time can be used as a smoothed estimate of the signal at the t-p+1 time.

We can clearly observe how the estimate is based on the entire previous signal and on p-1 following samples.

Since the least square error:

$$E\left[\left(s(\tau) - \hat{s}(\tau|t)\right)^2\right]$$

is a non-increasing function of the t-τ observation interval, advantage could be taken from the use of a smoother with a q observation interval higher than p-1.

All this can be easily implemented by re-using the Kalman filter given by the (13) and (14) equations, but redefining the matrices $A(t)$, $\varphi(t)$ and c present in (12), as follows:

$$A_q(t) = \begin{bmatrix} a_1(t) & \cdots & a_p(t) & 0 & \cdots & \cdots & 0 & 0 \\ 1 & \cdots & 0 & \vdots & \ddots & \ddots & \vdots & \vdots \\ \vdots & \ddots & \vdots & \vdots & \ddots & \ddots & \vdots & \vdots \\ 0 & \cdots & 1 & 0 & \cdots & \cdots & 0 & 0 \\ 0 & \cdots & \cdots & 0 & 1 & \cdots & \cdots & 0 & 0 \\ \vdots & \ddots & \ddots & \vdots & \vdots & \ddots & \ddots & \vdots & \vdots \\ \vdots & \ddots & \ddots & \vdots & \vdots & \ddots & \ddots & \vdots & \vdots \\ 0 & \cdots & \cdots & 0 & 0 & \cdots & \cdots & 1 & 0 \end{bmatrix}$$

$$\varphi_q(t) = \begin{bmatrix} s(t) \\ s(t-1) \\ \vdots \\ s(t-p) \\ s(t-p-1) \\ \vdots \\ s(t-q+1) \end{bmatrix} \qquad b_q(t) = \begin{bmatrix} 1 \\ 0 \\ \vdots \\ 0 \\ 0 \\ \vdots \\ 0 \end{bmatrix}$$

in which we obviously have to consider $q > p$.

By using the superfluous components $s(t-p)$, ..., $s(t-q+1)$, this non-minimum realisation of the AR process in an increased state space makes it possible to achieve a local *forwards/backwards* reconstruction of the data blocks irrevocably distorted by impulsive noises.

2.2.1.2   *Reconstruction of the damaged samples*

Considering the realisation obtained by increasing the state, we now analyse how it can be used for a better interpolation of the damaged audio signal portions. Since:

$$\hat{\varphi}_q(t \mid t) = \left[E[s(t) \mid Y(t)], \ldots, E[s(t-q+1) \mid Y(t)] \right]^T$$

with $Y(t) = \{y(1), \ldots, y(t)\}$, the smoothed estimate of the signal is obtained by:

$$\hat{s}(t-q+1 \mid t) = d_q^T \, \hat{\phi}_q (t \mid t)$$

in which d_q is the vector defined by:

$$d_q^T = [0, \ldots, 0, 1]$$

When the indicating function $d(t)$ (compare equation 15) satisfies $d(t-q+1) = 1$, the quantity $\hat{s}(t-q+1 \mid t)$ can be interpreted as the optimum reconstruction, in the sense of least squares, of the sample $s(t-q+1)$, based on the whole past and on q-1 future measures (except from those affected by impulsive noise).

We can prove that for a signal described by an AR model of p order, it is necessary and sufficient to have a block of p future samples free from impulsive noises, in order to guarantee the best reconstruction quality of the damaged fragment (Niedzwiecki and Cisowski, 1996).

The use of an observation interval $q>p$ also allows the algorithm to restore a signal corrupted by a series of impulsive noises having a duration up to q-p samples, a case which occurs for example in presence of a scratch.

2.2.1.3   *Parameters estimate*

The considerations previously put forward are based on the assumption that the time-varying coefficients are known. If this cannot be verified in real applications, as usually occurs, an algorithm based on the Kalman filter can be used to follow the variation of $\theta(t)$.

Suppose that the change of coefficients can be locally described by the following *random walk* model:

$$a_i(t+1) = a_i(t) + w_i(t) \qquad\qquad i=1,\ldots,p \qquad\qquad (16)$$

where $w(t)=[w1(t), \ldots, w_p(t)]^T$ is a white Gaussian vector, independent from $e(t)$:

$$w(t) \sim N(0, \sigma_w^2 I_p)$$

the diagonal form adopted for the covariance matrix of $w(t)$ indicates that the parameters vary *independently* from one another. The quantity σ_w^2 can be considered as an index which shows how much the coefficients are expected to vary .

Rewriting the (16) equation in a vector form

$$\theta(t+1) = \theta(t) + w(t) \tag{17}$$

$$s(t) = \varphi^T(t-1)\,\theta(t-1) + e(t-1) \tag{18}$$

The estimate of $\theta(t)$ can be considered as a standard filtering problem in the states space.

The only *non standard* characteristic is represented by the fact that $\varphi^T(t\text{-}1)$ depends on the data.

The correspondent Kalman filter takes the form:

$$
\begin{cases}
\hat{\theta}(t \mid t) = \hat{\theta}(t-1 \mid t-1) + L_\theta(t)\,\varepsilon_\theta(t) \\[6pt]
\Sigma_\theta(t \mid t) = \left(I_p - L_\theta(t)\varphi^T(t-1)\,\Sigma_\theta(t-1 \mid t-1) + \xi I_p\right) \\[6pt]
\varepsilon_\theta(t) = s(t) - \varphi^T(t-1)\,\hat{\theta}(t-1 \mid t-1) \\[6pt]
L_\theta(t) = \dfrac{\Sigma_\theta(t-1 \mid t-1)\varphi(t-1)}{\varphi^T(t-1)\,\Sigma_\theta(t-1 \mid t-1)\varphi(t-1)+1}
\end{cases}
\tag{19}
$$

where:

$$\xi = \frac{\sigma_w^2}{\sigma_e^2} \tag{20}$$

and $\Sigma\theta(t \mid t)$ is the covariance matrix of the estimate error, normalised in comparison with σ_e^2, in fact:

$$p\big(\theta(t) \mid S(t)\big) = \left(\hat{\theta}(t \mid t, \sigma_e^2\,\Sigma_\theta(t \mid t)\right)$$

where $S(t) = \{s(1), \dots, s(t)\}$.

The algorithm tracking properties depend on the choice of a single ξ parameter, that is on the relation between the parameters variation degree and the variance of signal innovation. This parameter must be chosen according to the *non-stationarity* degree of the system to identify.

2.2.1.4 Detection/tracking/restoration with EKF

The smoothing algorithm previously described and the tracking algorithm given in (19), if considered alone, do not solve the restoration problem in a completely adapting way. Actually in order to obtain the first filter, the vector $\theta(t)$ of the AR model coefficients was assumed to be known, whereas to derive the second, the hypothesis that $\varphi(t)$ was measurable (that is, that the non-noisy signal was available) had been made. The two procedures must be appropriately matched and must work parallely in order to simultaneously solve the problem of model identification and of audio signal cleaning.

Suppose now that the original signal $x(t)$ is corrupted by a mixture of broadband noise $z(t)$ and from impulsive disturbances $v(t)$. The measure $y(t)$ can be thought as:

$$y(t) = s(t) + z(t) + v(t) = b_q{}^T \, \varphi_q(t) + v(t) \tag{21}$$

where:

$$z(t) \sim N(0, \sigma_z{}^2), \quad \text{noise estimated}$$

$$v(t) \sim N(0, \sigma_v{}^2), \quad \text{impulsive noise} \quad \sigma_v{}^2 = \begin{cases} 0 & se \ d(t) = 0 \\ \infty & se \ d(t) \neq 0 \end{cases}$$

As previously explained, setting $\sigma_v{}^2 = \infty$, we indicate that the measure at the t time has no information regarding the original signal: the sample is simply considered missing.

Now combining the vector of the $\theta_p(t)$ parameters with that of the $\varphi_p(t)$ state in a state vector of dimension $p+q$ and $x(t) = \left[\varphi_q^T(t), \theta_p^T(t)\right]^T$ the equations (12), (16) and (21) can be rewritten as:

$$\begin{cases} x(t+1) = f[x(t)] + w(t) \\ y(t) = cTx(t) + \zeta(t) \end{cases} \tag{22}$$

where:

$$f[x(t)] = \begin{bmatrix} A_q(t) & 0 \\ 0 & I_p \end{bmatrix}, \quad \omega(t) = \begin{bmatrix} b_q e(t) \\ w(t) \end{bmatrix}, \quad \zeta(t) = z(t) + v(t) \tag{23}$$

and $c^T = \left[b_q^T, 0_p^T\right]$.

The problem of simultaneous parameters identification of the time-varying processing (estimation of $\theta_p(t)$) and of noise removal from the audio signal (estimation of $\varphi_q(t)$) can be considered as a non linear filtering problem in the space state. The transition matrix f depends in fact on the $x(t)$ state, since it also contains the AR model coefficients of the signal.

A sub-optimal solution to this problem is based on the Extended Kalman Filter (EKF) theory. Denote with $F(t)$ the state transition matrix of the *linearized* system, where:

$$F(t) = \nabla_x f[x]_{x=\hat{x}(t|t)} = \begin{bmatrix} A_q(t|t) & \hat{\varphi}_p^T(t|t) \\ 0_{p\times q} & 0_{q-1\times p} \\ & I_p \end{bmatrix} \tag{24}$$

where $A(t|t) = A[\hat{\theta}(t|t)]$, $\hat{\varphi}_p(t|t)$ indicates the vector constituted of the first p components $\hat{\varphi}(t|t)$ and:

$$\hat{x}(t|t) = \begin{bmatrix} \hat{\varphi}_q(t|t) \\ \hat{\theta}_q(t|t) \end{bmatrix}$$

is the trajectory of the filtered state obtained from the EKF

Consider then:

$$\Omega = cov[\omega(t)]/\sigma_e^2 = \begin{bmatrix} b_q b_q^T & 0 \\ 0 & \xi I_p \end{bmatrix} \tag{25}$$

where $\xi = \sigma_w^2/\sigma_e^2$.

The extended Kalman filter equations for the system governed by (22) are therefore (Anderson and Moore, 1979):

- **Prediction phase:**

$$\begin{cases} \hat{x}(t\,|\,t-1) = f[\hat{x}(t-1\,|\,t-1)] \\ \Sigma(t\,|\,t-1) = F(t-1)\,\Sigma(t-1\,|\,t-1)\,F(t-1)^T + \Omega \end{cases} \tag{26}$$

- **Updating phase:**

$$\begin{cases} \hat{x}(t\,|\,t) = \hat{x}(t\,|\,t-1) + L(t)\,\varepsilon(t) \\ \Sigma(t\,|\,t) = \left(I_{p+q} - L(t)\,c^T\right)\Sigma(t\,|\,t-1) \end{cases} \tag{27}$$

where $\varepsilon(t)$ indicates the prediction error $\varepsilon(t) = y(t) - c^T \hat{x}(t\,|\,t-1)$ and $L(t)$ is the Kalman gain whose value depends on the impulses detector $\hat{d}(t)$:

$$L(t) = \begin{cases} \dfrac{\Sigma(t\,|\,t-1)\,c}{c^T\,\Sigma(t\,|\,t-1)\,c + k_0} & \hat{d}(t) = 0 \\ 0 & \hat{d}(t) \neq 0 \end{cases}$$

$$k_0 = \sigma_z^2/\sigma_e^2$$

The filter can be initialised with the following values:

$$\hat{x}(0\,|\,0) = 0 \quad \Sigma(0\,|\,0) = \begin{bmatrix} 0 & 0 \\ 0 & \delta I_p \end{bmatrix} \tag{28}$$

where δ is a positive constant of high value (~ 100) to take into account that nothing concerning $\theta(0)$ is known a priori.

2.2.1.5 Impulsive noise detection
A simple impulsive noise detector is given by the well-known rule of the $\mu \times \sigma$ product used in statistics.

The value assumed by the prediction error $\varepsilon(t)$ (the Kalman filter innovation) is to be checked from time to time:

$$d(t) = \begin{cases} 0 & se\ |\varepsilon(t)| \leq \mu \hat{\sigma}_\varepsilon(t) \\ 1 & se\ |\varepsilon(t)| > \mu \hat{\sigma}_\varepsilon(t) \end{cases} \tag{29}$$

in which $\hat{\sigma}_\varepsilon(t)$ is calculated starting from an estimate of the variance $\hat{\sigma}_e(t)$ of $e(t)$ (the signal innovation):

$$\hat{\sigma}_\varepsilon(t) = \eta(t)\,\hat{\sigma}_e(t-1) \tag{30}$$

$$\eta(t) = c^T\, \Sigma(t\,|\,t-1)\,c + k_0 \tag{31}$$

and μ is a parameter depending on the user and determining the threshold for impulsive noise detection.
$\hat{\sigma}_\varepsilon(t)$ is the local maximum likelihood (exponentially weighted) of the signal innovation variance for the calculation of which the following equation is used:

$$\hat{\sigma}_e(t) = \begin{cases} \lambda\hat{\sigma}_e^{\,2}(t-1) + (1-\lambda)\dfrac{\varepsilon^2(t)}{\eta(t)} & se\ \hat{d}(t) = 0 \\ \hat{\sigma}_e^{\,2}(t-1) & se\ \hat{d}(t) \neq 0 \end{cases} \tag{32}$$

with λ ($0 << \lambda < 1$) determining the adaptation speed of the filter to the signal variations.

Note that at every time the decision threshold of the detector depends on the $\eta(t)$ multiplier which is updated by the EKF.

2.2.1.6 *Split version of the EKF*

The advantages deriving from the appropriate matching of the two filters, which are integrated in the above-illustrated EKF, are evident above all in presence of impulsive disturbances. In particular, since the tracking routine uses the previously filtered and reconstructed signal, the samples classified as disturbances can no longer be used during the parametric estimate process.

Although the algorithm is elegant and efficient and meets the needs of a unified and really actually adaptive approach to the restoration problem, it has however the disadvantage of being heavy from the point of view of computation. Furthermore, the filter integrated structure prevents from independently modifying the *tracking* and *restoration* parts which make it up, with the purpose, for example, of optimising the performances.

Therefore a variant of the previously described EKF, capable of solving such problems, guaranteeing higher flexibility and efficiency (Niedzwiecki, 1997), is herein presented.

From the complete filter, two parts, so to say, will be *retrieved*:

- one for signal filtering;
- one for parameters identification.

The two components will be then appropriately interconnected so as to guarantee a sort of *equivalence* with the starting EKF. The strategy consists in using the signal previously *cleaned* from the filtering routine for the parametric estimate and to employ the estimates so obtained to continue the processing.

2.2.1.7 *Signal Filtering*

In order to isolate the part concerning signal restoration from the identification part, the vector of the parameters in the algorithm described by the equations (26) and (27) is considered known. More precisely we assume $\theta(t-1|t-1) \equiv \theta(t-1)$ and $\xi=0$.

Under these hypothesis we have:

$$\Sigma(t-1|t-1) = \begin{bmatrix} \Sigma_\varphi(t-1|t-1) & 0 \\ 0 & 0 \end{bmatrix}$$

which combined with (26) and (27) gives:

$$\Sigma(t \,|\, t-1) = \begin{bmatrix} \Sigma_\varphi(t \,|\, t-1) & 0 \\ 0 & 0 \end{bmatrix}$$

where:

$$\Sigma_\varphi(t \,|\, t-1) = A_q(t-1 \,|\, t-1)\, \Sigma_\varphi(t-1 \,|\, t-1)\, A_q{}^T(t-1 \,|\, t-1) + b_q b_q{}^T$$

and the a *posteriori* covariance matrix becomes:

$$\Sigma_\varphi(t \,|\, t) = \left(I_q - L_\varphi(t)\, b_q{}^T \right) \Sigma_\varphi(t \,|\, t-1)$$

with

$$L_\varphi(t) = \frac{\Sigma_\varphi(t \,|\, t-1)\, b_q}{b_q{}^T \Sigma_\varphi(t \,|\, t-1)\, b_q + k(t)} \qquad (33)$$

$$k(t) = var\left[\zeta(t) \right] / \sigma_e{}^2$$

and $\zeta(t)$ given by (23).

    Ultimately the filter for the signal filtering alone, obtained by (26) and (27) considering the model parameters known, is described by the following equations:

■ **Prediction phase**

$$\begin{cases} \hat{\varphi}_q(t \,|\, t-1) = A_q(t-1 \,|\, t-1)\hat{\varphi}_q(t \,|\, t-1) \\ \Sigma_\varphi(t \,|\, t-1) = A_q(t-1 \,|\, t-1)\, \Sigma_\varphi(t-1 \,|\, t-1)\, A_q{}^T(t-1 \,|\, t-1) + b_q b_q{}^T \end{cases} \qquad (34.a)$$

■ **Updating phase**

$$\begin{cases} \hat{\varphi}_q(t \,|\, t) = \hat{\varphi}_q(t \,|\, t-1) + L_\varphi(t)\, \varepsilon(t) \\ \Sigma_\varphi(t \,|\, t) = \left(I_q - L_\varphi(t)\, b_q{}^T \right) \Sigma_\varphi(t \,|\, t-1) \end{cases} \qquad (34.b)$$

where:

$$\varepsilon(t) = y(t) - \hat{\theta}_p^{\ T}(t-1|t-1)\hat{\varphi}_p(t-1|t-1) \tag{35}$$

and:

$$L_\varphi(t) = \begin{cases} \dfrac{\Sigma_\varphi(t|t-1)b_q}{b_q^{\ T}\Sigma_\varphi(t|t-1)b_q + k_0} & \text{se } \hat{d}(t) = 0 \\[4mm] 0 & \text{se } \hat{d}(t) \neq 0 \end{cases} \tag{36}$$

The scale factor $\eta(t)$ necessary to the (30) and (31) equations for the calculation of the impulsive noise detection threshold is obtained from

$$\eta(t) = b_q^{\ T}\Sigma_\varphi(t|t-1)b_q + k_0 \tag{37}$$

Note as in absence of impulsive noise ($\hat{d} \equiv 0$) the gain $L_\varphi(t)$ coincides with (13) and corresponds to that of a classic Kalman filter planned for a system with known coefficients $\theta(t) \equiv \hat{\theta}(t|t)$.

2.2.1.8 *Parameters estimate*

Suppose now to ignore the fact that the measures are affected by noise. This is the same as considering that in the equations (26)-(27) we have $\hat{\varphi}_q(t-1|t-1) \equiv \varphi_q(t-1)$ and $k(t)=0$. Under these hypothesis we obtain:

$$\Sigma(t-1|t-1) = \begin{bmatrix} 0 & 0 \\ 0 & \Sigma_\theta(t-1|t-1) \end{bmatrix}$$

Further mathematical passages (Niedzwiecki, 1997) show that in this case

$$\Sigma(t|t-1) = \begin{bmatrix} \Sigma_\varphi(t|t-1) & \Sigma_{\varphi\theta}(t|t-1) \\ \Sigma_{\theta\varphi}(t|t-1) & \Sigma_\theta(t|t-1) \end{bmatrix}$$

where:

$$\Sigma_\varphi(t\,|\,t-1) = \begin{bmatrix} \varphi_p^{\,T}\,(t-1\,|\,t-1)\,\psi_p\,(t-1\,|\,t-1)+1 & 0 \\ 0 & 0 \end{bmatrix}$$

$$\Sigma_{\theta\varphi}(t\,|\,t-1) = \begin{bmatrix} \psi_p\,(t-1\,|\,t-1) & 0 \end{bmatrix}$$

$$\Sigma_{\varphi\theta}(t\,|\,t-1) = \Sigma_{\theta\varphi}^{\,T}\,(t\,|\,t-1)$$

$$\Sigma_\theta(t\,|\,t-1) = \Sigma_\theta(t-1\,|\,t-1) + \xi I_p$$

and

$$\psi_p(t-1\,|\,t-1) = \Sigma_\theta(t-1\,|\,t-1)\hat{\varphi}_p(t-1\,|\,t-1)$$

Furthermore we have:

$$\Sigma(t\,|\,t) = \begin{bmatrix} 0 & 0 \\ 0 & \Sigma_\theta(t\,|\,t) \end{bmatrix}$$

where:

$$\Sigma_\theta(t\,|\,t) = \Sigma_\theta(t\,|\,t-1) - L_\theta(t)\psi_p^{\,T}(t-1\,|\,t-1)$$

and

$$L_\theta(t) = \begin{cases} \dfrac{\psi_p(t-1\,|\,t-1)}{\hat{\varphi}_p^{\,T}(t-1\,|\,t-1)\,\psi_p(t-1\,|\,t-1)+1} & \text{se } \hat{d}(t)=0 \\ 0 & \text{se } \hat{d}(t) \neq 0 \end{cases} \tag{38}$$

The parameter identification procedure can be therefore retrieved from the EKF and expressed in the compact formula:

$$\begin{cases} \hat{\theta}_p(t\,|\,t) = \hat{\theta}_p(t-1\,|\,t-1) + L_\theta(t)\,\varepsilon(t) \\ \Sigma_\theta(t\,|\,t) = \left(I_p - L_\theta(t)\hat{\varphi}_p^{\,T}(t-1\,|\,t-1)\right)\Sigma_\theta(t-1\,|\,t-1) + \xi I_p \end{cases} \tag{39}$$

Note as the innovation is still given by (35):

$$\varepsilon(t) = y(t) - \hat{\theta}_p^{\,T}(t-1\,|\,t-1)\hat{\varphi}_p(t-1\,|\,t-1)$$

and is the *same* for both Kalman filters.

It is necessary to bear in mind that, beyond appearances, the method above-described remains in substance an *equivalent* (but more favourable) method through which to carry out the complete EKF calculations, whose filtering/tracking characteristics and good properties are maintained (Niedzwiecki, 1997). Thus, during each stage the two parts exchange the information about the *state vectors* they estimate: that of the filtered samples $\varphi(t)$ and that of the coefficients $\theta(t)$ respectively

The important aspect of this EKF version is its modularity: in fact each component can be modified without jeopardising the good working of the other.

2.2.1.9 *EKF with variable order*

As previously acknowledged, high values for the q parameter guarantee a better removal of long impulsive noises series (typical for example of scratches). However it is evident how all this also implies a higher computational weight, that is approximately of q^2 order in relation to the specific structure of the state transition matrix $A_q(t)$.

Exploiting the modularity of the Split-EKF a change of its filtering part, capable of using a variable observation interval q (Niedzwiecki, 1997), is hereafter proposed.

Denote with $\Sigma_q(t\,|\,t-1)$ and with $\Sigma_q(t\,|\,t)$ the matrices of $q{\times}q$ dimensions previously defined with the notation $\Sigma_\varphi(t\,|\,t-1)$ and $\Sigma_\varphi(t\,|\,t)$ respectively.

The algorithm starts setting $q = p$ and is divided into four fundamental steps, which are:

- **1st step** – The Kalman filter order is increased from q to $q+1$ and the *prediction* phase is carried out.

$$\hat{\varphi}_{q+1}(t \,|\, t-1) = \begin{bmatrix} \hat{\theta}_p(t-1\,|\,t-1)\hat{\varphi}_p(t-1\,|\,t-1) \\ \hat{\varphi}_q(t-1\,|\,t-1) \end{bmatrix}$$

$$\Sigma_{q+1}(t \,|\, t-1) = \begin{bmatrix} \alpha(t\,|\,t-1) & \beta^T(t\,|\,t-1) \\ \beta(t\,|\,t-1) & \Sigma_q(t-1\,|\,t-1) \end{bmatrix}$$

where

$$\alpha(t\,|\,t-1) = \hat{\theta}_p^{\,T}(t-1\,|\,t-1)\,\Sigma_p(t-1\,|\,t-1)\,\hat{\theta}_p(t-1\,|\,t-1) + 1$$

$$\beta(t\,|\,t-1) = \Sigma p(t-1\,|\,t-1)\,\hat{\theta}_q(t-1\,|\,t-1)$$

$\hat{\theta}_q(t-1\,|\,t-1)$ denote the vector $\hat{\theta}_p(t-1\,|\,t-1)$ extended with q-p zeros:

$$\theta_q(t-1\,|\,t-1) = \begin{bmatrix} \hat{\theta}_p(t-1\,|\,t-1) \\ 0 \end{bmatrix}$$

and $\Sigma_p(t\text{-}1\,|\,t\text{-}1)$ is the principal sub matrix $(p{\times}p)$ of $\Sigma_q(t\text{-}1\,|\,t\text{-}1)$.

- **2nd step** – After having calculated $\varepsilon(t)$ through the (35) equation and $k(t)$ with the (33), the presence of impulsive noises in the last p stages is controlled by using an index of the type:

$$D(t) = \sum_{i=1}^{p} \hat{d}(t-i+1) \tag{40}$$

in which the value of $\hat{d}(t)$ is obtained through the detector (29) by using the (37) in replacement of the (31).

- **3rd step** – The *updating* phase is calculated.

$$\begin{cases} \hat{\phi}_{q+1}(t\,|\,t) = \hat{\phi}_{q+1}(t\,|\,t-1) + L_{q+1}(t)\,\varepsilon(t) \\ \Sigma_{q+1}(t\,|\,t) = \left(I_{q+1} - L_{q+1}(t)\,b_{q+1}{}^{T}\right)\Sigma_{q+1}(t\,|\,t-1) \end{cases}$$

in which the value of the Kalman gain is obtained by (36).

- **4th step** – If $D(t){\neq}0$ and $q{<}q_{max}$ we go back to the **1st step**. Otherwise:
  - The reconstructed signal components are retrieved

$$\hat{s}(t-i+1\,|\,t) = d_i\,\hat{\phi}_{q+1}(t\,|\,t)\text{, with } i{=}p{+}1,\ldots,q{+}1$$

  - the Kalman filter order is reduced from $q{+}1$ to p removing the last $(q{-}p{+}1)$ elements of $\hat{\phi}_{q+1}(t\,|\,t)$ and the last $(q{-}p{+}1)$ columns and lines of $\Sigma_{q+1}(t\,|\,t)$ to respectively form $\hat{\phi}_p(t\,|\,t)$ e $\Sigma_p(t\,|\,t)$.

We return therefore to the **2nd step**. Note that if there are no impulsive noise signals for an interval p samples long, $q{=}p$ is obtained and therefore the order of the smoothing filter is equivalent to its nominal value $p{+}1$. On the contrary, in presence of impulsive noise, the dimension of the filter state is gradually increased up to the point in which $D(t){=}0$ (or rather at least p future measures are free from impulsive noises and therefore *reliable*) or till q reaches the maximum value q_{max} (a sort of *safety valve* which in the experimentations carried out was set to value equal to 100).

The part of the EKF regarding the identification of the model parameters remains unchanged. It is important to pay attention to the fact that the order of the AR model used for the signal processing remains however *fixed* and equal to p. It is the state space dimension of the filtering routine only which is increased. By exploiting this non-minimality of representation a sort of buffer is obtained, which allows the filter to wait until a sample block (equal to p) is available, in which no impulsive noises are signalled.

The algorithm so modified implies a significant gain regarding the calculation time, since in every moment it allows for the use of the minimum order possible for the filter, without in any way jeopardising the restoration quality.

2.2.1.10 *Scaling of the filter.*

A serious (but often ignored) problem affecting the parametric identification algorithms based on the Kalman filter is their sensitivity to scale changes of the signal.

After having chosen an adequate value for $\xi = \dfrac{\sigma_w^2}{\sigma_e^2}$ it occurs, in fact, that a possible multiplication of the measures for a factor $\Delta \neq 1$ modifies the result of the parameters estimate unless ξ is not reduced by Δ^2 times. In order to avoid this problem we can use the following *exponentially weighted least squares* algorithm (EWLS) in replacement of the (39) equations.

$$\psi_p(t-1|t-1) = \Sigma_\theta(t-1|t-1)\hat{\phi}_p(t-1|t-1)$$

$$L_\theta(t) = \begin{cases} \dfrac{\psi_p(t-1|t-1)}{\hat{\phi}_p^T(t-1|t-1)\psi_p(t-1|t-1)} & \text{se } \hat{d}(t) = 0 \\[2ex] 0 & \text{se } \hat{d}(t) \neq 0 \end{cases} \tag{41}$$

$$\begin{cases} \hat{\theta}_p(t|t) = \hat{\theta}_p(t-1|t-1) + L_\theta(t)\varepsilon(t) \\[1ex] \Sigma_\theta(t|t) = \dfrac{1}{\gamma}\Big[\Sigma_\theta(t-1|t-1) - L_\theta(t)\psi_p^T(t-1|t-1)\Big] \end{cases}$$

The effective memory of the EWLS filter can be easily controlled through the   forgetting-constant' γ, with $0 << \gamma < 1$. Differently from what happens in the (39) equations, this algorithm enjoys the property of being invariant to the scale changes of the input signal.

We can see that the (41) equation foresees to calculate the matrix Σ as difference between two matrices defined positive. This can lead to the destruction of the definitiveness characteristic because of numeric errors (due to the limited precision of the data in the calculators), especially in the longer processing.

A solution proposed in literature is known as *Joseph stabilized recursion* (Niedzwiecki, 2000). It includes a slightly different form for the last of the equations (41):

$$\Sigma_\theta(t \mid t) = \frac{1}{\gamma}\left[I_p - L_\theta(t)\hat{\phi}_p^{\ T}(t-1 \mid t-1)\right]\Sigma_\theta(t-1 \mid t-1)\left[I_p - L_\theta(t)\hat{\phi}_p^{\ T}(t-1 \mid t-1)\right] + L_\theta(t)L_\theta^{\ T}(t) \quad (42)$$

The (42) equation is mathematically equal to the last of the equations (41), but it guarantees to maintain the characteristics of having a defined filter covariance matrix.

To conclude, we mention other techniques through which to enhance the numeric strength and which guarantee a better calculation efficiency than that of (42). They are known as *square root filtering* and foresee to replace $\Sigma(t \mid t)$ with its square root, defined as:

$$\Sigma(t \mid t) = S(t \mid t)S^T(t \mid t) \quad\quad\quad (43)$$

In this way, non-positive covariance matrices will never be obtained. Numerous solutions are present in literature for the calculation of $S(t \mid t)$, see for example (Niedzwiecki, 2000).

2.2.1.11 *Tracking with variable memory.*

In this paragraph an innovative solution for filter tracking, and more precisely an adaptive implementation for the γ *constant*, that is the *forgetting constant*, will be examined. Remember that such a quantity approximately represents the memory filter, that is the weighing factor of the *remoter* samples (in the past) compared with the more recent one. The fundamental concept is that the memory of an adaptive filter should be in inverse relation to the *average information* of the new data acquired.

Starting from the previously presented EWLS filter, suppose to set the non fixed γ quantity, but in turn updated to each stage of the algorithm. The problem of how such a quantity has to be calculated arises.

In the article (Fortescue, Kershenbaum, e Ydstie, 1981) the authors suggest choosing $\gamma(t)$ so as to *stabilize* the measurement metric of the average information, defined as weighed sum of the estimate least squares errors:

$$q(t) = \sum_{i=0}^{t-1} w_t(i)\left(y(t-1) - \varphi^T(t-1)\hat{\theta}_p(t|t)\right)^2 \tag{44}$$

where $\hat{\theta}_p(t|t)$ is the vector of the coefficients estimated of the model, and in addition:

$$w_t(0) = 1$$
$$w_t(i) = \prod_{j=0}^{i-1} \gamma(t-j), \qquad 0 < i \le t-1 \tag{45}$$

are the weights to give to the past estimates. Note that the last relation can be written in a recursive way:

$$w_t(i) = \gamma(t)w_{t-1}(i-1) \quad 1 \le i \le t \tag{46}$$

In other words, if the estimate error is small, it means that a good model is available, or that the estimator is sensitive enough to efficiently follow the variation. In these cases a reasonable strategy consists in choosing a forgetting constant close to the unit (*long* memory). Vice versa, if the error is big, the estimator sensitiveness must be increased by choosing a smaller forgetting constant (*short* memory) till the error becomes small again. We propose setting:

$$q(t) = q(t-1) = q_0 \tag{47}$$

in this way we assure that the estimate process is always based on the same quantity of information.

Therefore, the following recurrence equation, which expresses the evolution of $q(t)$, can be solved compared with $\gamma(t)$:

$$
\begin{aligned}
q(t) &= \gamma(t)q(t-1) + \left[1 - \hat{\varphi}_p^{\,T}(t\,|\,t)\Sigma(t\,|\,t)\hat{\varphi}_p(t\,|\,t)\right]\varepsilon^2(t) \\
&= \gamma(t)q(t-1) + \left[1 - L_\theta^{\,T}(t)\hat{\varphi}_p(t\,|\,t)\right]\varepsilon^2(t) \\
&= \gamma(t)q(t-1) + \frac{\gamma(t)\varepsilon^2(t)}{\gamma(t) + \hat{\varphi}_p^{\,T}(t\,|\,t)\Sigma(t-1\,|\,t-1)\hat{\varphi}_p(t\,|\,t)}
\end{aligned}
\tag{48}
$$

Replacing (47) in (48) and solving it, we obtain :

$$
q_0 = \gamma(t)q_0 + \frac{\gamma(t)\varepsilon^2(t)}{\gamma(t) + \beta(t)}
\tag{49}
$$

where we set:

$$
\beta(t) = \hat{\varphi}_p^{\,T}(t|t)\Sigma t|\,-1|t-1\,)\,\hat{\varphi}_p(t|t)
\tag{50}
$$

and the explicit expression for the calculation of $\gamma(t)$ is therefore:

$$
\gamma(t) = \frac{\delta(t) + \sqrt{\delta^2(t) + 4\beta(t)}}{2}
\tag{51}
$$

where:

$$
\delta(t) = 1 - \beta(t) - \frac{\varepsilon^2(t)}{q_0}
\tag{52}
$$

In the article (Ydstie and Sargent, 1986) the authors suggest approximating the exact solution (51) by setting $\gamma(t)+\beta(t) \cong 1+\beta(t)$ and ultimately obtaining:

$$\gamma(t) = \frac{1}{1+\dfrac{\varepsilon^2(t)}{(1+\beta(t))q_0}} \tag{53}$$

In order to obtain an estimate of the quantity order of the parameter q_0, consider the following reflection. In the hypothesis of a time-varying system and with a constant memory the following relations are valid:

$$\theta(t)=\theta, \qquad \gamma(t)=\gamma_0, \qquad 1-\gamma_0 <<1 \tag{54}$$

from which:

$$\hat{\theta}_p(t \,|\, t) \cong \theta \tag{55}$$

and:

$$y(t-i) - \varphi^T(t-i)\hat{\theta}_p(t \,|\, t) \cong z(t-i) \tag{56}$$

where $z(i)$ is the measure noise. In this case we can calculate:

$$\lim_{t\to\infty} E[q(t)] \cong k_0 \sigma_z^2 = q_0, \qquad k_0 = \sum_{i=0}^{\infty}(\gamma_0)^i = \frac{1}{1-\gamma_0} \tag{57}$$

or, in a more compact form

$$q_0 = \frac{1}{1-\gamma_0}\sigma_z^2 \tag{58}$$

It is also important to foresee a minimum threshold for the forgetting constant so as to limit the reactivity of the tracking routine. Furthermore in order to make the pace of $\gamma(t)$ less abrupt, ε^2 is replaced with the average value of the last M values (new parameter of the algorithm) of the prediction square error in (53), that is:

$$\varepsilon^2(t) \leftarrow \frac{1}{M} \sum_{i=0}^{M-1} \varepsilon^2(t-i) \tag{59}$$

Finally, the tracking part for the algorithm VFF becomes:

$$\beta(t) = \hat{\varphi}_p^{\mathrm{T}}(t \mid t)\Sigma(t-1 \mid t-1)\hat{\varphi}_p(t \mid t)$$

$$\gamma(t) = max \left[\gamma_{min}, \frac{1}{1 + \dfrac{\dfrac{1}{M}\sum_{i=0}^{M-1}\varepsilon^2(t-i)}{(1+\beta(t))q_0}} \right]$$

$$L_\theta(t) = \begin{cases} \dfrac{\Sigma(t-1 \mid t-1)\hat{\varphi}_p(t-1 \mid t-1)}{\beta(t)+\gamma(t)} & \text{if } \hat{d}(t)=0 \\ 0 & \text{if } \hat{d}(t) \neq 0 \end{cases} \tag{60}$$

$$\Sigma_\theta(t \mid t) = \frac{1}{\gamma(t)}\left[I_p - L_\theta(t)\hat{\varphi}_p^{T}(t-1 \mid t-1) \right]\Sigma_\theta(t-1 \mid t-1)$$

These formulas are devised to replace the (41) equations. Note that being a derivation of the EWLS, there is still the invariance property compared with the scale change of the signal. In addition, the problem concerning the calculation of the differences of matrices defined positive is no longer present and there is therefore no need for stabilized versions.

The fact that this VFF algorithm is heuristic and does not guarantee the protection of the Σ matrix divergence is to be emphasised; however we can prove that if the components of $\hat{\varphi}_p(t \mid t)$ are non null and *linearly dependent*, the forgetting constant is never forced to assume the value 1.

2.2.1.12 Dead zone

Paradoxically, in many cases problems concerning the identification of the coefficients can arise when the estimate algorithm performances of the parameters become too *good* (Niedzwiecki, 2000). The prediction error or a measure correlated to such prediction can be considered as an indicator of the tracking quality. Problems of numerical nature arise when the following condition takes place:

$$\varepsilon(t) \to 0 \tag{61}$$

We could therefore consider interrupting the parameter estimate phase in case of a prediction error under a certain threshold δ

- if $|\varepsilon(t)| \leq \delta > 0$:

$$\begin{cases} \hat{\theta}_p(t \mid t) = \hat{\theta}_p(t-1 \mid t-1) \\ \Sigma_\theta(t \mid t) = \Sigma_\theta(t-1 \mid t-1) \end{cases} \tag{62}$$

- Otherwise we proceed with the tracking phase seen in the previous paragraphs.

Although this solution is very promising from the theoretical point of view, it nonetheless presents the problem regarding the determination of the δ threshold. A solution based on the a priori knowledge of the measure noise variance is proposed:

$$\delta = \eta \sigma_z \tag{63}$$

with the parameter $\eta \in [1,2]$.

2.2.1.13 Decimation of the tracking part

We have previously seen that the main purpose of the EKF split version is the possibility to optimize the performances in terms of calculation capacity by exploiting their modularity. We propose an optimisation which does not present a purely theoretical justification. The idea is simple: instead of updating the signal model at every stage of the algorithm we can do it every two or four stages. In this way we can improve efficiency since the processing of the split 2 (among which the stability control) are carried out in time in a decimated way. This method is not totally orthodox for the following reason: the equivalence of the split version to the standard version is no longer guaranteed. In fact, the latter foresees the recursion at every stage in order to update the model. In addition, it is no longer true that for the split 2 the innovation is calculated still with the (35) equation , that is as that of the split 1. In particular, if we continue to use the (35) equation, we will obtain values for ε smaller than their real measure. A slowdown of the algorithm convergence must be therefore expected.

2.2.1.14 *Alarm clustering*

During the experimentations, it occurred that in some cases the restoration was carried out in a more reliable way, if the impulsive disturbances detection was forced to form clusters made up of consecutive sequences of *one* (up to a maximum of q_{max}-p-1) alternated to at least p *zeros*.

This method prevents from erroneously accepting as good a sample localized between two samples affected by impulsive noise.

The practical realisation is obtained by calculating other two indexes after the equation (40):

$$D1(t) = \sum_{i=1}^{cw1} \hat{d}(t-i+1) \quad 0 \le cw1 < p$$

$$D2(t) = \sum_{i=1}^{cw2} \hat{d}(t-i+1) \quad 0 \le cw2 < p$$

in which $\hat{d}(t)$ is obtained always through the detector described using the (37) instead of the equation (31).

The *cw*1 and *cw*2 parameters determine the amplitude of the windows used for the *clustering* in the two sections (*filtering* and *tracking*) of the EKF

The values obtained for $D1(t)$ and $D2(t)$ will be then used in replacement of $\hat{d}(t)$ in (36) and (38) (in 42 if the EWLS is used or in (60) if the VFF for the tracking part is used) respectively.

2.2.1.15 *Filter parameter choice*

In order to obtain a good working of the filter, carrying out a priori a correct choice of the value of some parameters which determine the dynamic is necessary. First of all, the p order of the AR model to use is to be decided. Generally speaking, a high value of p should improve the restoration quality in virtue of a higher number of future samples used and of a model theoretically more faithful to the signal . However we have to bear in mind that the computational complexity of the algorithm can be considered in first approximation proportional to the square of the *total* order of the EKF (that is, to the $q+p$ dimension of the state vector $x(t)=[\varphi_q^T(t), \theta_p^T(t)]$) and that such a value increases if impulsive disturbances are detected. Furthermore, in presence of background noise, excessively high orders for the model can easily cause problems concerning the coefficients identification, without producing improvements in the output SNR. On the basis of these considerations and of different tests carried out on heterogeneous audio material (synthetic examples, musical signals) for the model we chose to use order values included between 8 and 16 in relation to the signal typology on which we intend to intervene. We can assume the choice $p = 12$ a compromise between quality needs and computational problems.

As mentioned above, the tracking capacities of the standard algorithm (39) are regulated by the value of $\xi=\sigma_w^2/\sigma_e^2$, which connects the input noise variance of the *random walk* (the model of the coefficients variation in time) to that of the signal innovation. In fact ξ can be considered as a sort of time-variance indicator of the signal. In the practical experimentations (with unitary bottom-scale signal) choosing $\xi =10^{-5}$ usually guaranteed good performances. Higher values (10^{-4}) can cause, in some cases, stability problems in the estimate of the model (the coefficients tend to vary too much). On the contrary, lower values (10^{-6} and less) can bring about an excessive slowness in following the signal variations.

In case we use the EWLS for the tracking part the problem concerning the choice of ξ does not exist, but it is replaced by that of finding an appropriate value for the *forgetting constant* γ . The experimentation and analysis about the routine behaviour lead us to identify in $[1\text{-}10^{-5}, 1\text{-}10^{-6}]$ the interval to which we can refer to for the value of γ.

As far as the VFF algorithm parameters choice is concerned, a value in the interval $[1\text{-}10^{-4}, 1\text{-}10^{-5}]$ can be accepted as a variation field of the threshold for the forgetting constant (γ_{min}). On the basis of the experimentations carried out, $[0.1, 10.0]$ was the useful variation field of the q_0 parameter.

We also have to mention the choice of λ in (32), a parameter which governs the updating of the local estimate of σ_e^2 inside the impulse detector. In the tests carried out, a value included between 0.97 and 0.98 was always used without the arising of particular problems.

More critical is the choice of the threshold μ used for click detection in (29). For example, assuming that the ε innovation of the filter is Gaussian, a value of $\mu=3$ (corresponding to the classic rule of '3-sigma') is to be considered very aggressive, since it could generate a false alarm almost every 400 samples (that is, every 0.01 second if a frequency of 44.1 kHz is used for the sampling of the signal to restore). The possibility of running into false click signals is to be taken in high regard since it can lead to disastrous consequences in the restoration process. The fact that in presence of an impulsive disturbance the filter reacts setting at zero the Kalman gain temporally, until the alarms cease (or also a bit further if a value different from 0 is used for the clustering windows) is to be duly taken into consideration. The cancellation of $L(t)$ opens up the feed-back chain and cuts the a posteriori filter updating part, which goes on only in virtue of the model available at that moment, which is also *frozen* until normality is restored. In this circumstance the EKF trusts the model completely and interrupts the measures used. After only a small number of samples (which make up the click) the signal is expected to behave again in a *reasonable* way and the filter is expected to refasten. If, on one hand, this mechanism guarantees sturdiness, since all the corrupted samples are removed (which however would not give any information), on the other hand we infer how an excess of false signals can hinder the correct working of the system. The filter, in fact, could be blocked even more often thereby losing its capacity to adapt to the signal generating further false alarms in an avalanche propagation.

Note that some little irregularities in the innovation trend can be corrected in a *natural* way also by the *smoother*, without further inconvenience. Excessively high values of μ can cause a non adequate removal of the impulsive disturbances, which reveals itself with the presence of annoying residues in the restoring.

In order to give a quantity order we can think that $\mu \in [4, 6]$, however this topic will be analysed in detail in the following pages.

To conclude the choice of the *clustering*, windows amplitude is mentioned. Remember that *cw* represents the number of the samples expected after the last click signal before allowing the refastening to the measures of the two filter sections. It is clear that setting this value to zero can determine the acceptance of non completely reliable samples, above all if high values for μ are being used. *cw* choices close to p can, on the other hand, cause effects similar to those caused by a false alarm excess, in particular if the signal presents many clicks and low values for μ are being used.

The availability of two windows (cw_1 and cw_2) allows for a higher flexibility of use, thereby making it possible to independently modify, on the basis of the above mentioned considerations, the behaviour of the two EKF sections depending on the will to favour either aggressiveness or sturdiness. As a rule, the values included between 2 and 4 are to be considered the norm, as they guarantee a good compromise between sturdiness and performances.

2.2.1.16 *Algorithm's blocks scheme*

Figure 7 shows a simplified algorithm blocks scheme which illustrates how starting from the noisy measures $y(t)$ and having at its disposal the necessary a priori knowledge on the noise $\hat{\sigma}_z^2$, the EKF carries out the signal filtering. The filter dynamic is however controlled by the detector of impulsive disturbances which, if necessary, intervenes modifying the Kalman gain. In particular, we wish to highlight how the statistic properties (the whiteness) of the prediction error $\varepsilon(t)$ are exploited in order to carry out this operation correctly.

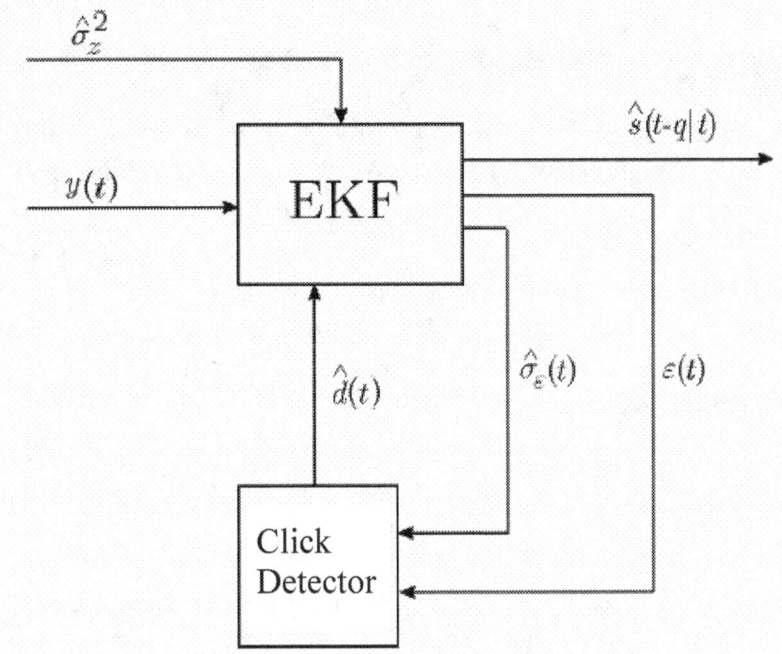

Fig. 7: *EKF Algorithm blocks scheme*

2.2.2   *Monte Carlo method with Rao-Blackwell strategy*

2.2.2.1 *State estimate*

When using signals generated by a physical system, as in the audio case, the model which regulates the temporal evolution can rarely be known in a deterministic way. It is therefore necessary to recur to a probabilistic description of the dynamics of the system.

Useful in this sense can be the use of the Markov process (also known as Markov chains in the case of numeric processes), in which the current state determines the evolution of the system in a context of aleatory indeterminable influences. In other words, the future states probability density is what is completely determined by the current state of the system. When the current state of a Markov process is not directly available, but only one of its functions is, we have a hidden Markov chain

As explained in (12), by using a description in the state space of the dynamic system, which can be represented with a hidden Markov process, it is possible to write:

$$\begin{cases} x_{t+1} = F(x_t) + R_t \\ y_t = G(x_t) + S_t \end{cases} \tag{64}$$

where x_t is a vector of d dimension representing the non-observable states of the system at the t time, y_t is a vector of the m dimension representing the observations made at the t time, F is a function in $\mathfrak{R}^d$, G is a function from $\mathfrak{R}^d$ to $\mathfrak{R}^m$, R_t is a vector of independent aleatory variables and with identical probability density and diagonal covariance matrix $R=\sigma_r^2 I$, as well as S_t is a vector of independent aleatory variables and with identical probability density with diagonal covariance matrix $S=\sigma_s^2 I$.

Therefore we have:

$$\begin{cases} x_{t+1} \sim f(x_{t+1} | x_t) \\ y_t \sim g(y_t | x_t) \end{cases} \tag{65}$$

that is x_{t+1} and y_t are not specified deterministically but they are aleatory vectors with $f(.|.)$ and $g(.|.)$ respectively state transition density functions and observation density functions depending on the probability density of R_t and S_t.

In the typical case in which R_t is a vector of aleatory variables with Gaussian probability density with R covariance matrix, the state transition density function becomes:

$$f(x_{t+1} | x_t) = N(F(x_t); R) =$$

$$= \frac{1}{\sqrt{(2\pi)^d \cdot det(R)}} \cdot exp\left(-\frac{1}{2}(x_{t+1} - F(x_t)) \cdot R^{-1} \cdot (x_{t+1} - F(x_t))' \right) \tag{66}$$

Usually the problem typologies which we want to solve given a Markov process are two:

- state estimate: in every t time we want to go back to an estimate of the current state x_t having at disposal only the $y_{1:t}$ observations up to the current time (intending with the term $y_{1:t} = \{y_1, y_2, ..., y_t\}$). This problem is known in literature with the term *filtering* even if it has nothing to do with the spectral manipulations often correlated to the filtering concept;

- trajectory estimate: we have at disposal an entire temporal series of observations $y_{1:T}$ and we want to go back to the complete trajectory estimate of the $x_{1:T}$ state. In literature this problem is often called smoothing.

In both problem typologies we have at disposal only the observation and we want to obtain an estimate of the state, calculated as the result of a function $\hat{x}(y)$ named estimator.

The minimum variance estimator is that given by the conditioned expectation:

$$\hat{x}(y) = E[x \mid y](y) = \int x \cdot p(x \mid y)\, dx \tag{67}$$

in case of state estimate it is therefore necessary to update at every stage the probability density $p(x_t \mid y_{1:t})$.

This updating can be obtained by using the recursive procedure

$$p(x_{t+1} \mid y_{1:t}) = \int p(x_t \mid y_{1:t}) \cdot f(x_{t+1} \mid x_t)\, dx_t$$

$$p(x_{t+1} \mid y_{1:t+1}) = \frac{g(y_{t+1} \mid x_{t+1}) \cdot p(x_{t+1} \mid y_{1:t})}{p(y_{t+1} \mid y_{1:t})} \tag{68}$$

Similarly the trajectory estimate requires the knowledge of the probability density which can be obtained recursively through a shift backwards in the temporal sequence, that is, it can be obtained by proceeding from the estimate of the following sample to the preceding one (*backward*).

$$p(x_t \mid y_{1:T}) = \int p(x_{t+1} \mid y_{1:T}) \cdot \frac{p(x_t \mid y_{1:t}) \cdot f(x_{t+1} \mid x_t)}{p(x_{t+1} \mid y_{1:t})}\, dx_{t+1} \tag{69}$$

In practice these recursive procedures can be expressed in a closed formula only for linear models of (64) in which F is a matrix and R_t a vector of aleatory variables with Gaussian density, by using the Kalman filter.

In fact under these conditions it results:

$$p(x_t \mid y_{1:t}) = \frac{1}{\sqrt{(2\pi)^d \cdot det(\Sigma_t)}} \cdot exp\left(-\frac{1}{2}(x_t - \varphi_t) \cdot \Sigma_t^{-1} \cdot (x_t - \varphi_t)'\right), \tag{70}$$

with

$$\Sigma_t = \left(F \cdot \Sigma_{t-1} \cdot F' + R\right)^{-1} + G' \cdot S^{-1} \cdot G,$$
$$\varphi_t = F \cdot \varphi_{t-1} + \Sigma_t \cdot G' \cdot S^{-1}\left(y_t - G \cdot F \cdot \varphi_{t-1}\right) \tag{71}$$

If there is no linearity or Gaussian condition of the model, an analytic expression for the density functions required does not exist. The classical approach in this case consists in using the Extended Kalman Filter, previously presented, in which the model is linearized and the traditional Kalman filter is applied to its linearization.

The problems connected to the Extended Kalman filter are due to the fact that the linearization not always implies a good approximation of the same, as in the case of highly non-Gaussian or multimodal functions and that the result is strictly linked to the initial conditions imposed and which not always are sufficiently correct.

An alternative strategy which can be used when the classic Kalman filter does not work is the Monte Carlo method.

The traditional Monte Carlo methods were planned to evaluate the result of some integrals, which can be intended as the expectation value of an aleatory quantity; in the typical case: $\int f(x) \cdot p(x)\, dx$, with $f(.)$ a function and $p(.)$ a probability density.

The idea at the basis of the Monte Carlo simulation consists in generating a high number (N) of independent aleatory variables called *particles* $x^{(1)}, x^{(2)}, \ldots, x^{(N)}$ having the probability density desired $p(x)$, for which the value of the integral can be approximated as:

$$\int f(x) \cdot p(x)\, dx \cong \frac{1}{N} \sum_{i=1}^{N} f(x^{(i)}) \tag{72}$$

In case we want to go back to the state estimate of a hidden Markov process, it is necessary to use the function (67) and the probability density which has to be respected when generating the N aleatory variables is $p(x_t|y_{1:t})$, which can be approximated with

$$p(x_t \mid y_{1:t}) \cong \sum_{i=1}^{N} w_t^{(i)} \cdot \delta(x_t - x_t^{(i)}) \tag{73}$$

where $\delta(.)$ is the Dirac Delta function and $w_t^{(i)}$ is a weight positive factor associated to the $x_t^{(i)}$ variable, which is calculated so as to obtain $\sum_{i=1}^{N} w_t^{(i)} = 1$

With this approximation for the a posteriori density, the integral value can be easily calculated with the approximation:

$$\hat{x}_t = \int f(x_t) \cdot p(x_t \mid y_{1:t}) \, dx_t \cong \sum_{i=1}^{N} f(x_t^{(i)}) \cdot w_t^{(i)} \tag{74}$$

Similarly, in case we want to obtain a trajectory estimate of a Markov process state (smoothing), N independent aleatory variables distributed according to the probability density $p(x_t|y_{1:T})$ have to be generated.

In the following pages the model adopted for the audio signal study consisting in a non-linear parameterization of an autoregressive time-varying (TVAR) model will be illustrated. Then the classic Monte Carlo algorithms for filtering and smoothing will be illustrated and analysed. Ultimately, an algorithm which exploiting the characteristics of the model used for the signal to enable performances superior to the classic algorithms will be described.

2.2.2.2 Audio model

To represent the audio signal the model mostly used is, as already illustrated in (1), an autoregressive model (AR).

In the simplest autoregressive models the coefficients associated to the previous samples are assumed constant during all the time interval considered. However to have a more realistic representation it is necessary to allow the coefficients to vary in time according a pre-established dynamic (or according to a dynamic established beforehand). In this way we obtain a time-varying autoregressive model (TVAR). In Vermaak, Andrieu, Doucet, and Godsill (1999) a TVAR model with stochastically variable parameters (*random walk*) is proposed and shows how this model turns out to be better than others in the context of vocal signals.

Following these indications we have decided to use a Gaussian evolution for the logarithm of the variance of the e_t white noise.

Defined $\phi_{e,t} = ln(\sigma^2_{e,t})$, we have:

$$f(\phi_{e,t} \mid \phi_{e,t-1}, \sigma^2_{\phi_e}) = N(\mu_{\phi_t}, \sigma^2_{\phi_e}) \tag{75}$$

with $\mu_{\phi_t} = ln(\alpha \cdot \sigma^2_{e,t-1})$ and α a coefficient just minor to 1 and therefore the updating of the value $\phi_{e,t}$ can be obtained easily by adding a Gaussian variable ξ_t with $\sigma^2_{\phi_e}$ to the previous value modified with the α parameter:

$$\phi_{e,t} = \mu_{\phi_{t-1}} + \xi_{t-1} \tag{76}$$

The simplest way to obtain a temporal variation of the $\{a_i\}_t$ coefficients consists in making them vary stochastically, by considering them as a vector of independent Gaussian aleatory variables with an average equal to the value at the previous instant and σ^2_a variance equal for every coefficient and considered constant during the entire time of analysis. The probability density associated to the coefficients is therefore:

$$f(a_{t,i} \mid a_{t-1,i}) = N(a_{t-1,i}, \sigma^2_a I_i) \tag{77}$$

and the updating of the coefficients is obtained by adding a Gaussian variable ζ_t to the previous value:

$$a_{t,i} = a_{t-1,i} + \zeta_{t-1} \tag{78}$$

The risk inherent in using such a random walk model is to obtain an instable system, which is a particularly dangerous condition for the restoration of an audio signal, since it can be further damaged instead of being improved.

For a system with parameters varying slowly in time, as in the audio case, the stability criterion is the same used for the common time-invariant systems: the roots of the associated polynomial:

$$1 - \sum_{i=1}^{p} a_{t,i} z^{-1} \qquad\qquad (79)$$

corresponding to the poles of the transferring function of the system expressed in the Zeta transform domain, must fall inside the unitary circle centred in the origin of the complex plane.

Instead of applying the random walk directly to the autoregressive model we preferred to re-parameterize the model in terms of reflection coefficients and apply the Gaussian temporal variance to them.

To pass from the coefficients of the autoregressive model to the reflection coefficients of the pylon model, also called partial correlation model (PARCOR), the recursive Levinson algorithm (Friedlander, 1982) was useed, which defines a non linear transformation, but which can be inverted between the $\{a_i\}$ coefficients and the $\{k_i\}$ reflection coefficients.

The procedure to obtain the $\{k_i\}$ starting from the $\{a_i\}$ coefficients is called 'Step Down' algorithm and can be schematized as follows:

- Define $a_j(i)$ the i-nth coefficient of a j order polynomial representing the AR model we want to segment.
- Initialize $k_p = a_p(p)$.
- For each j from p-1 to 1:

  - for each i from 1 to j define $a_j(i) = \dfrac{a_{j+1}(i) - k_{j+1} \cdot a_{j+1}(j-i+1)}{1 - k^2_{j+1}}$;

  - define $k_j = a_j(j)$.

Once the $\{k_{t,i}\}$ reflection coefficients valid at the t time have been found, the random walk is applied to them in order to find the reflection coefficients $\{k_{t+1,i}\}$ valid for the following time.

The probability density used is still Gaussian with the same σ^2_a variance for the direct coefficients:

$$f(k_{t+1,i} | k_{t,i}, \sigma^2_a) = N(k_{t,i}, \sigma^2_a I_i) \qquad\qquad (80)$$

The stability control for the reflection coefficients is simpler than that for the $\{a_{t,i}\}$ coefficients and requires the model to be stable when all the reflection coefficients $\{k_{t,i}\}$ are in module minor of the unity.

In the case in which at least one of the $\{k_i\}$ coefficients generated for the $t+1$ time results in module superior to 1 it is sufficient to repeat the procedure of random generation till the following condition takes place: $max\ \{|k_{t+1,i}|\} < 1$, thus assuring the model stability.

Once we have obtained a stable model expressed in function of the reflection coefficients $\{k_{t+1,i}\}$ valid for the temporal time $t+1$, it is necessary to retreat to the equivalent autoregressive model by exploiting the Levinson 'Step Up':

- Initialize the recorsion by defining: $a_0(0)=1$.
- For every j from 0 to p-1:
  - for every i from 1 to j define: $a_{j+1}(i)=a_j(i)+k_{j+1}\cdot a_j(j-i+1)$;
  - obtain: $a_{j+1}(j+1)=k_{j+1}$.

Applying the random walk to the pylon model, instead of directly applying it to the autoregressive model, is made necessary by the fact that the former can be seen as the schematisation of a generation mechanism of the signal through an wave guide made up of two cylinders of different diameter, which can represent a discrete approximation for the physical process of the sound genesis in many instruments and for the vocal signal.

For the smoothing phase, where an inverse temporal advancing is necessary, the same coefficients, concerning the same temporal time considered and generated during the filtering phase (during the normal temporal advancing) are used. This choice is dictated by the fact that the shape of the human vocal tract, comparable to a pipe, varies in time rather slowly because of physical constraints and therefore the reflection coefficients turn out to be very similar (practically identical) both in the forwarding and backwarding phases.

The state vector x_t for the specified model is divided into two components $x_t = [\varphi_t,\ \theta_t]$, where φ_t is the vector containing the last p samples:

$$\varphi_t = [s_{t-1},\ s_{t-2},\ldots,\ s_{t-p}] \tag{81}$$

and θ_t is the vector containing the reflection coefficients of the time-varying pylon model and the logarithm of the TVAR model Gaussian excitation variance:

$$\theta_t = [k_{t,1}, k_{t,2}, \ldots, k_{t,p}, \phi_{et}] \tag{82}$$

The observation signal y_t available is assumed to be the original signal s_t with the addition of white Gaussian noise z_t with known variance σ^2_z:

$$y_t = s_t + z_t \tag{83}$$

Therefore it results that $g(y_t | x_t) = N(s_t, \sigma^2_z)$.

The parameters σ^2_a (variance of the Gaussian random walk applied to the reflection coefficients) and $\sigma^2_{\phi_e}$ (variance of the Gaussian random walk applied to the logarithm of the Gaussian excitation variance of the autoregressive model) are supposed to be known and constant during the whole temporal interval of the simulation.

The model used can be written in the form:

$$\begin{cases} \varphi_t = T(\theta_t) \cdot \varphi_{t-1} + H(\theta_t) \cdot \omega_t \\ y_t = Z(\theta_t) \cdot \varphi_t + G(\theta_t) \cdot \omega_t \end{cases} \tag{84}$$

with:

$$T = \begin{bmatrix} a_{t,1} & a_{t,2} & a_{t,3} & \cdots & a_{t,p} \\ 1 & 0 & 0 & \cdots & 0 \\ 0 & 1 & 0 & \cdots & 0 \\ \cdot & \cdot & \cdot & \cdot & \cdot \\ 0 & 0 & \cdots & 1 & 0 \end{bmatrix}, H = \begin{bmatrix} 1 & 0 \\ 0 & 0 \\ \cdot & \cdot \\ 0 & 0 \end{bmatrix}, Z = \begin{bmatrix} 1 & 0 & \cdot & 0 \end{bmatrix}, G = \begin{bmatrix} 0 & 0 \\ 0 & 1 \end{bmatrix}$$

and:

$$\omega_t = \begin{bmatrix} e_t \\ z_t \end{bmatrix}$$

vector of the Gaussian processes concerning respectively the model excitation and the noise overlapping with the original signal, with covariance matrix:

$$Q_t = \begin{bmatrix} \sigma^2_{et} & 0 \\ 0 & \sigma^2_z \end{bmatrix}$$

2.2.2.3 Rao-Blackwell Strategy

One of the biggest problems connected to the Monte Carlo methods for the filtering and smoothing of a signal is the huge quantity of data created, data which are to be memorised and processed during the filtering and *smoothing*. In some cases, however, the model used allows us to establish a functional dependence between the variables belonging to the state. If what Doucet, De Freitas, Murphy, and Russel (2000) define "treatable substructures" exist, it is possible to define a function $h(.)$ which connects two or more state variables $x_1 = h(x_2, x_3, \ldots)$, we can marginalize the dependent variable conditioning it in comparison with the other state variables from which it depends. This strategy, known as "Rao-bleckwellisation" (Casella and Robert, 1996), is a technique for variance reduction generally used in statistics.

In the case of the audio signal studied with the proposed model, the state vector φ_t concerning the signal can be written as depending from the state vector θ_t: $\varphi_t = T(\theta_t) \cdot \varphi_{t-1} + H(\theta_t) \cdot \omega_t$, and therefore the only uncertainty degree associated with it, once θ_p is known, is given by the overlapped Gaussian noise $H(\theta_t) \cdot \omega_t$.

So the variable φ_t can be expressed with a Gaussian linear system for which it is possible and convenient to use the Kalman filter and the decomposition of the prediction error (Anderson and Moore, 1979), (Harvey, 1989), (Bari *et al.*, 2001).

2.2.2.4 Monte Carlo Filtering with the Rao-Blackwell strategy.

In order to obtain an estimation of the current state x_t knowing the observation $y_{1:t}$ consider the expectation:

$$E[x_t \mid y_{1:t}] = \int x_t \cdot p(x_t \mid y_{1:t}) \, dx_t . \tag{85}$$

By using the division of the state $x_t = [\varphi_t, \theta_t]$ we obtain:

$$p(x_t \mid y_{1:t}) = p(\varphi_t, \theta_t \mid y_{1:t}) = p(\varphi_t \mid \theta_t, y_{1:t}) \cdot p(\theta_t \mid y_{1:t}) \qquad (86)$$

The first factor of the decomposition $p(\varphi_t \mid \theta_t, y_{1:t})$ is a Gaussian density and therefore the integration connected to it will be carried out by means of Kalman filter, whereas for the integration concerning the second factor $p(\theta_t \mid y_{1:t})$ the Monte Carlo Method will be used. The latter, exploiting Bayes's Law can be written as:

$$p(\theta_t \mid y_{1:t}) = \frac{p(y_t \mid \theta_t, y_{1:t-1}) \cdot p(\theta_t \mid y_{1:t-1})}{p(y_t \mid y_{1:t-1})} \qquad (87)$$

quantity which results proportional to the integral:

$$p(\theta_t \mid y_{1:t}) \propto \int p(y_t \mid \theta_{1:t}, y_{1:t-1}) \cdot p(\theta_{1:t-1} \mid y_{1:t-1}) \cdot f(\theta_t \mid \theta_{t-1}) \, d\theta_{1:t-1} \qquad (88)$$

Supposing that from the previous stage we have at disposal the approximation generated with the Monte Carlo method for $p(\theta_{1:t-1} \mid y_{1:t-1})$, it is possible to generate N samples $\theta^{(i)}_t$ by using the knowledge of $f(\theta_t \mid \theta_{t-1})$.

We can therefore write the approximation:

$$p(\theta_t \mid y_{1:t}) \cong \sum_{i=1}^{N} w^{(i)}_t \cdot \delta(\theta_t - \theta^{(i)}_t) \qquad (89)$$

where $w^{(i)}_t$ is a weight associated to each sample, proportional to the density which can be calculated a posteriori by using the Kalman Filter and the decomposition of the prediction error:

$$w^{(i)}_t \propto p(y_t \mid \theta^{(i)}_{1:t}, y_{1:t-1}) \qquad (90)$$

Combining the equations concerning the Kalman filter (71) with the filter which uses the Monte Carlo Method we obtain the following algorithm:

Filtering Algorithm with the Monte Carlo Method by applying the Rao-Blackwell strategy:

▪ Define $f(\theta_1 | \theta_0) = f(\theta_1)$ and initialize $\{\varphi_{1|0}, \Sigma_{1|0}\}$ with $\{\varphi_{1|1}, \Sigma_{1|1}\}$, $\varphi_{1|1}$ can be easily set to a value equal to that of the null vector, while $\Sigma_{1|1}$ is a diagonal matrix $k \cdot I$, with k as a high value (for example $k=100$) to have a high initial uncertainty. Using consistent values is generally convenient for the initialisation of the parameters θ_1. We used an algorithm similar to that which is being illustrated, with the only purpose of constituting a *tracking* procedure of the model coefficients, instead of obtaining an estimation of the complete state. By properly regulating the parameters associated with the variance of the random walks for the coefficients σ^2_a and for the logarithm of the Gaussian excitation variance of the AR model $\sigma^2_{\phi_e}$, we can obtain the convergence towards plausible values by using less than 2000 samples.

▪ For each t from 1 to T:

- For each i from 1 to N generate samples for the values of the parameters $\theta^{(i)}_t$ according to the density $f(\theta_t | \theta^{(i)}_{t-1})$;

- For each realisation of $\theta^{(i)}_{1:t}$ s update the statistic knowledge of φ_t using the Kalman filter:

  ➤ update the matrix $T(\theta^{(i)}_t)$ and calculate the covariance matrix $Q(\theta^{(i)}_t)$;

  ➤ calculate with Kalman the quantities concerning φt in order to define:

  $$p(\varphi_t | \theta_{1:t}, y_{1:t-1}) = N(\varphi^{(i)}_{t|t-1}, \Sigma^{(i)}_{t|t-1}),$$

  $$p(\varphi_t | \theta_{1:t}, y_{1:t}) = N(\varphi^{(i)}_{t|t}, \Sigma^{(i)}_{t|t}),$$

  where:

$$\begin{cases} \varphi^{(i)}_{t|t-1} = T(\theta^{(i)}_t) \cdot \varphi^{(i)}_{t-1|t-1} \\ \Sigma^{(i)}_{t|t-1} = T(\theta^{(i)}_t) \cdot \Sigma^{(i)}_{t-1|t-1} \cdot T(\theta^{(i)}_t)' + H \cdot Q(\theta^{(i)}_t) \cdot H' \end{cases} \qquad \text{(91.a)}$$

$$\begin{cases} \varphi^{(i)}{}_{t|t} = \varphi^{(i)}{}_{t|t-1} + \Sigma^{(i)}{}_{t|t-1} \cdot Z' \cdot F^{(i)}{}_t{}^{-1} \cdot (y_t - y^{(i)}{}_{t|t-1}) \\ \Sigma^{(i)}{}_{t|t} = \Sigma^{(i)}{}_{t|t-1} \cdot (I - Z' \cdot F^{(i)}{}_t{}^{-1} \cdot Z \cdot \Sigma^{(i)}{}_{t|t-1}) \end{cases} \qquad (91.b)$$

with $F^{(i)}{}_t$ covariance matrix $F^{(i)}{}_t = cov(y_t, y_t | y^{(i)}{}_{1:t-1}, \theta^{(i)}{}_{1:t})$, computable through the prediction error decomposition:

$$F^{(i)}{}_t = Z \cdot \Sigma_{t|t-1} \cdot Z' + G \cdot Q(\theta^{(i)}{}_t) \cdot G' \qquad (92.a)$$

and with $y^{(i)}{}_{t|t-1}$ expectation $y^{(i)}{}_{t|t-1} = E[y_t | y^{(i)}{}_{1:t-1}, \theta^{(i)}{}_{1:t}]$, computable as:

$$y^{(i)}{}_{t|t-1} = Z \cdot \varphi^{(i)}{}_{t|t-1} ; \qquad (92.b)$$

- for each i from 1 to N value the weight coefficient d $w^{(i)}{}_t$ which is to associate with each $\theta^{(i)}{}_t$ by using the prediction error decomposition (2.42.a and 2.42.b):

$$w^{(i)}{}_t \propto N(y_t - y^{(i)}{}_{t|t-1}, F^{(i)}{}_t) =$$

$$= \frac{1}{\sqrt{2 \pi F^{(i)}{}_t}} \cdot exp\left(-\frac{1}{2 F^{(i)}{}_t} \cdot (y_t - y^{(i)}{}_{t|t-1})\right); \qquad (93)$$

- normalize the $w^{(i)}{}_t$ calculated so that $\sum_{i=1}^{N} w^{(i)}{}_t = 1$;

- re-sample $\{\theta^{(i)}{}_t, \varphi^{(i)}{}_{t|t}, \Sigma^{(i)}{}_{t|t}\}$ N times by using the above-described deterministic reselection algorithm.

2.2.2.5 Monte Carlo Smoothing with Rao-Blackwell strategy.
As anticipated in the paragraph concerning smoothing, the estimation of the whole trajectory of the state $x_{1:T}$ can be calculated with less uncertainty if we dispose of the whole observation signal evolution $y_{1:T}$.

The results obtained by employing the Monte Carlo filtering with the above-described Rao-Blackwell strategy can be used to create an estimator of the trajectory (*smoother*). We want to generate an estimation of $x_{1:T}$ by producing samples which respect the state density $p(x_{1:T} \mid y_{1:T})$.

It is possible to factorize such density as follows:

$$
\begin{aligned}
p(x_{1:T} \mid y_{1:T}) &= p(\varphi_{1:T}, \theta_{1:T} \mid y_{1:T}) = \\
&= p(\varphi_T, \vartheta_T \mid y_{1:T}) \cdot \prod_{t=1}^{T-1} p(\varphi_t, \theta_t \mid \varphi_{t+1:T}, \theta_{t+1:T}, y_{1:T})
\end{aligned}
\tag{94}
$$

The topic of the product can be written as:

$$
\begin{aligned}
p(\varphi_t, \theta_t \mid \varphi_{t+1:T}, \theta_{t+1:T}, y_{1:T}) &= \int p(\varphi_t, \theta_{1:t} \mid \varphi_{t+1:T}, \theta_{t+1:T}, y_{1:T}) d\theta_{1:t-1} = \\
&= \int p(\theta_{1:t} \mid \varphi_{t+1:T}, \theta_{t+1:T}, y_{1:T}) \cdot p(\varphi_t \mid \varphi_{t+1:T}, \theta_{1:T}, y_{1:T}) d\theta_{1:t-1}
\end{aligned}
\tag{95}
$$

By exploiting the property of the Markov processes, the density concerning the trajectory of the parameters can be written as follows:

$$
p(\theta_{1:t} \mid \varphi_{t+1:T}, \theta_{t+1:T}, y_{1:T}) = p(\theta_{1:t} \mid \varphi_{t+1}, \theta_{t+1}, y_{1:T}).
\tag{96}
$$

By applying Bayes's rule to the second element and marginalizing it compared to $\theta_{1:t}$ we obtain

$$
\begin{aligned}
p(\theta_{1:t} \mid \varphi_{t+1:T}, \theta_{t+1:T}, y_{1:T}) &= \frac{p(\varphi_{t+1}, \theta_{t+1} \mid \theta_{1:t}, y_{1:t}) \cdot p(\theta_{1:t} \mid y_{1:t})}{\int p(\varphi_{t+1}, \theta_{t+1} \mid \theta_{1:t}, y_{1:t}) \cdot p(\theta_{1:t} \mid y_{1:t}) d\theta_{1:t}} = \\
&= \frac{p(\varphi_{t+1} \mid \theta_{1:t+1}, y_{1:t}) \cdot f(\theta_{t+1} \mid \theta_t) \cdot p(\theta_{1:t} \mid y_{1:t})}{\int p(\varphi_{t+1} \mid \theta_{1:t+1}, y_{1:t}) \cdot f(\theta_{t+1} \mid \theta_t) \cdot p(\theta_{1:t} \mid y_{1:t}) d\theta_{1:t}}
\end{aligned}
\tag{97}
$$

By using the approximation generated in the previous filtering cycle for the density $p(\theta_{1:t} \mid y_{1:t}) \cong \sum_{i=1}^{N} w^{(i)}{}_t \cdot \delta(\theta_{1:t} - \theta^{(i)}{}_{1:t})$ it is possible to write:

$$p(\theta_{1:t} \mid \varphi_{t+1:T}, \theta_{t+1:T}, y_{1:T}) \cong$$

$$\cong \sum_{i=1}^{N} \left(\frac{w^{(i)}_t \cdot p(\varphi_{t+1} \mid \theta^{(i)}_{1:t}, \theta_{t+1}, y_{1:t}) \cdot f(\theta_{t+1} \mid \theta^{(i)}_t)}{\sum_{j=1}^{N} w^{(j)}_t \cdot p(\varphi_{t+1} \mid \theta^{(j)}_{1:t}, \theta_{t+1}, y_{1:t}) \cdot f(\theta_{t+1} \mid \theta^{(j)}_t)} \right) \times \qquad (98)$$

$$\times \delta(\theta_{1:t} - \theta^{(i)}_{1:t}) = \sum_{i=1}^{N} w^{(i)}_{t|t+1} \cdot \delta(\theta_{1:t} - \theta^{(i)}_{1:t})$$

with:

$$w^{(i)}_{t|t+1} \propto w^{(i)}_t \cdot p(\varphi_{t+1} \mid \theta^{(i)}_{1:t}, \theta_{t+1}, y_{1:t}) \cdot f(\theta_{t+1} \mid \theta^{(i)}_t) \qquad (99)$$

The equation (95) can therefore be rewritten as:

$$p(\varphi_t, \theta_t \mid \varphi_{t+1:T}, \theta_{t+1:T}, y_{1:T}) \cong$$

$$\cong \sum_{i=1}^{N} w^{(i)}_{t|t+1} \cdot p(\varphi_t \mid \varphi_{t+1:T}, \theta^{(i)}_{1:t}, \theta_{t+1:T}, y_{1:T}) \cdot \delta(\theta_t - \theta^{(i)}_t) \qquad (100)$$

Henceforth we explain how it is possible to generate realisations of the trajectory $\{\tilde{\varphi}_{1:T}, \tilde{\theta}_{1:T}\}$ by using as our reference probability density the approximation (100) obtained.

Supposing to know an estimation of the trajectory portion $\{\tilde{\varphi}_{t+1:T}, \tilde{\theta}_{t+1:T}\}$, we generate an estimation for $\tilde{\theta}_t$ by using as our reference density the approximation (100) similarly to what is illustrated in the paragraph concerning generic smoothing with the Monte Carlo method:

$$\tilde{\theta}_{1:t} \sim \sum_{i=1}^{N} w^{(i)}_{t|t+1} \cdot \delta(\theta_{1:t} - \theta^{(i)}_{1:t}) \qquad (101)$$

Actually it is sufficient to select a $\tilde{\theta}_{1:t} = \theta^{(j)}_{1:t}$ between the N realisations $\theta^{(i)}_{1:t}$ generated in the previous filtering cycle. The choice has to be carried out according to the weight $w^{(i)}_{t|t+1}$ which is calculated for each N possibility $\theta^{(i)}_{1:t}$.

Using the AR model hypothesis with Gaussian excitation (1), the density $p(\varphi_{t+1} \mid \theta^{(i)}_{1:t}, \theta_{t+1}, y_{1:t})$ present in the computation of the quantity $w^{(i)}_{t|t+1}$ (99) becomes a multidimensional Gaussian density with mean $\varphi^{(i)}_{t+1|t}$ and covariance $\Sigma^{(i)}_{t+1|t}$, quantities which are easy to calculate by using the Kalman filter equations (91.a) as well as the density $f(\widetilde{\theta}_{t+1} \mid \theta^{(i)}_t)$ with mean $\widetilde{\theta}_{t+1}$ and covariance matrix is a multidimensional Gaussian density.

$$
C = \begin{bmatrix}
\sigma^2_a & 0 & . & 0 & 0 \\
0 & \sigma^2_a & . & 0 & 0 \\
. & . & . & . & . \\
0 & 0 & . & \sigma^2_a & 0 \\
0 & 0 & . & 0 & \sigma^2_{\varphi e}
\end{bmatrix}.
$$

The weight constant to calculate for each realisation $\theta^{(i)}_{1:t}$ is therefore:

$$
w^{(i)}_{t|t+1} \propto w^{(i)}_t \cdot \frac{1}{\sqrt{(2\pi)^P \cdot det(\Sigma^{(i)}_{t+1|t})}} \times
$$

$$
\times exp\left(-\frac{1}{2}(\widetilde{\varphi}_{t+1} - \varphi^{(i)}_{t+1|t}) \cdot \Sigma^{(i)-1}_{t+1|t} \cdot (\widetilde{\varphi}_{t+1} - \varphi^{(i)}_{t+1|t})'\right) \times \qquad (102)
$$

$$
\times \frac{1}{\sqrt{(2\pi)^{p+1} \cdot det(C)}} \cdot exp\left(-\frac{1}{2}(\widetilde{\theta}_{t+1} - \theta^{(i)}_t) \cdot C^{-1} \cdot (\widetilde{\theta}_{t+1} - \theta^{(i)}_t)'\right).
$$

We chose to use the selection strategy illustrated in the paragraph concerning generic smoothing, limiting the choice to the more probable $\theta^{(i)}_{1:t}$ and therefore with a bigger associated weight in order to repeat the procedure many times, thus obtaining different realizations of the trajectory which are not all equal among them.

Once the value of the parameters $\theta^{(j)}_{1:t}$ is selected (and hence also the values $\widetilde{\varphi}_{t|t} = \varphi^{(j)}_{t|t}$, $\widetilde{\Sigma}_{t|t} = \Sigma^{(j)}_{t|t}$), it is possible to calculate the density $p(\varphi_t \mid \theta^{(j)}_{1:t}, \widetilde{\theta}_{t+1:T}, \varphi_{t+1:T}, y_{1:T})$, which is simply, even on the basis of the hypothesis put forward on the model used, a multidimensional Gaussian density with mean $\widetilde{\varphi}_{t|T}$ and covariance matrix o $\widetilde{\Sigma}_{t|T}$:

$$p(\varphi_t \mid \theta^{(j)}_{1:t}, \widetilde{\theta}_{t+1:T}, \widetilde{\varphi}_{t+1:T}, y_{1:T}) = N(\widetilde{\varphi}_{t|T}, \widetilde{\Sigma}_{t|T}) \tag{103}$$

where $\widetilde{\varphi}_{t|T}$ e $\widetilde{\Sigma}_{t|T}$ can be calculated by using the Kalman equations concerning temporal advancing in inverse sense.

$$\widetilde{\varphi}_{t|T} = \widetilde{\varphi}_{t|t} + S_t \cdot \left(\widetilde{\varphi}_{t+1|T} - T(\widetilde{\theta}_{t+1}) \cdot \widetilde{\varphi}_{t|t} \right)$$
$$\widetilde{\Sigma}_{t|T} = \widetilde{\Sigma}_{t|t} + S_t \cdot \left(\widetilde{\Sigma}_{t+1|T} - \widetilde{\Sigma}_{t+1|t} \right) \cdot S_t' \tag{104}$$

with $S_t = \widetilde{\Sigma}_{t|t} \cdot T(\widetilde{\theta}_{t+1})' \cdot \widetilde{\Sigma}^{-1}_{t+1|t}$.

These equations use quantities available, which do not necessarily need to be recalculated, if during the previous filtering cycle they had been memorised (with a significant memory waste, but allowing for a higher execution speed).
Repeating M times the generation process:

$$\widetilde{\theta}_t \sim \sum_{i=1}^{N} w^{(i)}_{t|t+1} \cdot \delta(\theta_t - \theta^{(i)}_t)$$
$$\widetilde{\varphi}_t \sim N(\widetilde{\varphi}_{t|T}, \widetilde{\Sigma}_{t|T}) \tag{105}$$

and proceeding with a temporal evolution in an inverse sense from T to 1, it is possible to generate M different estimations of the trajectory $\{\widetilde{\varphi}_{1:T}, \widetilde{\theta}_{1:T}\}$ which reflect the probability density $p(\varphi_{1:T}, \theta_{1:T} \mid y_{1:T})$.

Henceforth the algorithm developed for the trajectory estimation of the Markov system state, which models the audio signal, to be restored is presented.

Smoothing algorithm with the Monte Carlo method by applying the Rao-Blackwell strategy:

- Carry out a filtering cycle with the Monte Carlo method with the Rao-Blackwell strategy as presented in the previous paragraph, thus generating for each t from 1 to T the N quantities: $\{w^{(i)}_t,\ \theta^{(i)}_t,\ \varphi^{(i)}_{t|t},\ \Sigma^{(i)}_{t|t}\}$.

- Chose $\tilde{\theta}_T = \theta^{(j)}_T$ with $w^{(j)}_T$ probability and initialize: $\tilde{\varphi}_{T|T} = \varphi^{(j)}_{T|T}$, $\tilde{\Sigma}_{T|T} = \Sigma^{(j)}_{T|T}$.

- The $\tilde{\varphi}_T$ estimation is generated creating samples according to the Gaussian density $\tilde{\varphi}_T \sim N(\tilde{\varphi}_{T|T}, \tilde{\Sigma}_{T|T})$.

- For each t from T-1 to 1:
  - for each i from 1 to N calculate the quantities $\varphi^{(i)}_{t+1|t}$ e $\Sigma^{(i)}_{t+1|t}$ by using the equation (91.a) of the Kalman filter;
  - for each i from 1 to N calculate the weight $w^{(i)}_{t+1|t}$ associated with each realisation $\{\varphi^{(i)}_{t+1|t}$ e $\Sigma^{(i)}_{t+1|t}\}$ by using the formula:
  $$w^{(i)}_{t|t+1} \propto w^{(i)}_t \cdot N(\varphi^{(i)}_{t+1|t}, \Sigma^{(i)}_{t+1|t}) \cdot f(\tilde{\theta}_{t+1} | \theta^{(i)}_t);$$
  - normalize the weights $w^{(i)}_{t|t+1}$ so that $\sum_{i=1}^{N} w^{(i)}_{t|t+1} = 1$;
  - select an index j among the indexes i=1:N on the basis of the normalized weights $w^{(i)}_{t|t+1}$ with the selection algorithm presented in the paragraph concerning the generic smoothing and define $\tilde{\theta}_t = \theta^{(j)}_t$;
  - define $\{\tilde{\varphi}_{t|t}, \tilde{\Sigma}_{t|t}, \tilde{\Sigma}_{t+1|t}\} = \{\varphi^{(j)}_{t|t}, \Sigma^{(j)}_{t|t}, \Sigma^{(j)}_{t+1|t}\}$ and apply the Kalman equations (104) to obtain $\tilde{\varphi}_{t|T}$ e $\tilde{\Sigma}_{t|T}$;
  - generate the estimation $\tilde{\varphi}_t$ according to the Gaussian density $\tilde{\varphi}_t \sim N(\tilde{\varphi}_{t|T}, \tilde{\Sigma}_{t|T})$.

Repeating the procedure M times we can obtain M independent estimations of the trajectory of the state $\tilde{x}_{1:T} = [\tilde{\varphi}_{1:T}, \tilde{\theta}_{1:T}]$.

Considering the operation $\tilde{y}_t = Z \cdot \tilde{\varphi}_t$ which provides the first component of the vector $\tilde{\varphi}_t$, we obtain M estimations of the original signal, cleaned from a consistent part of the overlapped white Gaussian noise. With a simple mean operation on the M estimations we obtain an audio restoration whose characteristics in terms of signal to noise ratio (SNR) are presented in the chapter concerning validation.

2.2.2.6 Monte Carlo smoothing with the Rao-Blackwell strategy with blocks division.
Filtering with the Monte Carlo method presented is an efficient algorithm, thanks to its intrinsically parallel structure which can be exploited for computational purposes.

Furthermore, the samples generated at the current temporal instant depend only on that of the previous instant and therefore, there is no need to memorize a long temporal series of samples in a filtering context. On the contrary, when we estimate the trajectory with the Monte Carlo smoothing the information concerning the complete temporal evolution of the samples generated need to be memorized. Therefore it is a method which cannot be applied on very lengthy temporal series, such as audio signals of discrete quality, unless we drastically reduce the number of samples generated, with consequent algorithm performance losses. To avoid this type of problem, it is necessary to subdivide the signal into small blocks and apply the method to each block, paying careful attention to memorize the data needed to pass from one block to the other.

The starting temporal series is therefore divided in R non-overlapped blocks, with T as the length of each of them in terms of number of samples. Actually the length can be different from one block to the other but, for convenience's sake, we choose a common T equal to 2000 samples.

We apply a filtering cycle to each block, thus generating the samples and the pertaining weights: $\left\{ x^{(i)}{}_t, w^{(i)}{}_t \right\}$. At the beginning of every new block the filtering algorithm is initialized with the final samples and the pertaining weights generated in the previous block. A smoothing analysis is then applied to each block, thus generating M estimations of the clean trajectory limited to the block being examined.

The connection of the estimations obtained by the analysis of every single block supplies M estimations of the total trajectory, which after having been mediated, spawns an estimation of the starting total signal.

2.2.2.7 Filter parameter choice
In order to use the smoothing algorithm with the Monte Carlo method described in the previous chapter, it is necessary to determine the value of numerous parameters used during the processing and connected both to

the model employed for the signal and to the method itself. The treated signal is assumed normalised so that its module is minor to the unity.

We have to decide the p order of the TVAR model used, that is, the number of the past samples which determine the value of the current samples. As a rule of thumb, a high value of p should improve the restoration quality, in virtue of a higher number of samples used and of a model theoretically more faithful to the signal. However we have to bear in mind that the computational complexity of the algorithm significantly depends on the order of the model used and that very high orders do not imply surely better results in terms of output signal to noise ratio SNR.

On the basis of the considerations put forward and on the laboratory tests carried out, we decided to use values included between 6 and 12 according to the signal type on which intervention is needed. We can consider that the choice $p=9$ constitutes a good compromise.

A second parameter which has to be fixed is the variance (σ^2_a) of the Gaussian random walk with which the coefficients are varied to obtain a temporal variation of the model. Actually there is no need to apply the *random-walk* to the coefficients, since this procedure is already implicit in the Monte Carlo method with the Rao-Blackwell strategy. In fact this procedure foresees that at every stage N different values are generated for each coefficient of the model. This functionality is carried out summing quantities generated according the pre-established probability density $f(\theta_t | \theta_{t-1})$ to the coefficients of the previous stage. In the implementation developed we chose to use a Gaussian probability density with mean θ_{t-1} and variance σ^2_{mc}. It is therefore evident how the application of a Gaussian random walk to obtain a time-varying model can be included in the generation of N coefficients, due to the Monte Carlo method. It is sufficient to use a value which takes into account both of the requirements for the variance σ^2_{mc}. In order to obtain a sufficiently efficient trajectory estimator the trajectories generated must be different and non-degenerating in a single trajectory. If we want to obtain good results it is therefore necessary to use a probability density with a rather high variance. In the tests carried out we observed that a value of the variance σ^2_{mc} sufficient to obtain trajectories different from one another can vary between 10^{-3} and 10^{-5} depending on the signal being considered. For these values the time-variance of the model is fully respected, without needing to add a further random walk to the coefficients.

A random walk which must be considered is that connected to the variance of the noise used as input in the signal model. In fact we decided to develop the logarithm of the noise variance according to a Gaussian walk: $\phi_{et} = ln(\sigma^2_{et})$. As outlined above, the Gaussian density used is $f(\phi_{et} | \phi_{et-1}, \sigma^2_{\phi}) = N(\mu_{\phi t}, \sigma^2_{\phi e})$, with $\mu_{\phi t} = ln(\alpha \cdot \sigma^2_{et-1})$.

For α we decided to fix a value minor to the unity, but very close to 1. In fact we noticed that the whole procedure of filtering with the Monte Carlo Method presents an automatic behaviour of the parameters setting up, since the selection carried out between the N samples tends to favour the one generated by using the best parameters. The temporal evolution of the variance σ^2_{et}, due only to the selection carried out on the samples without using the corrective constant α, is characterized by a spontaneous progressive reduction , as shown in Figure 8. Therefore, there is no need to use an excessively low value for the constant α, and the position $\alpha=0.9998$ turned out to be a good choice.

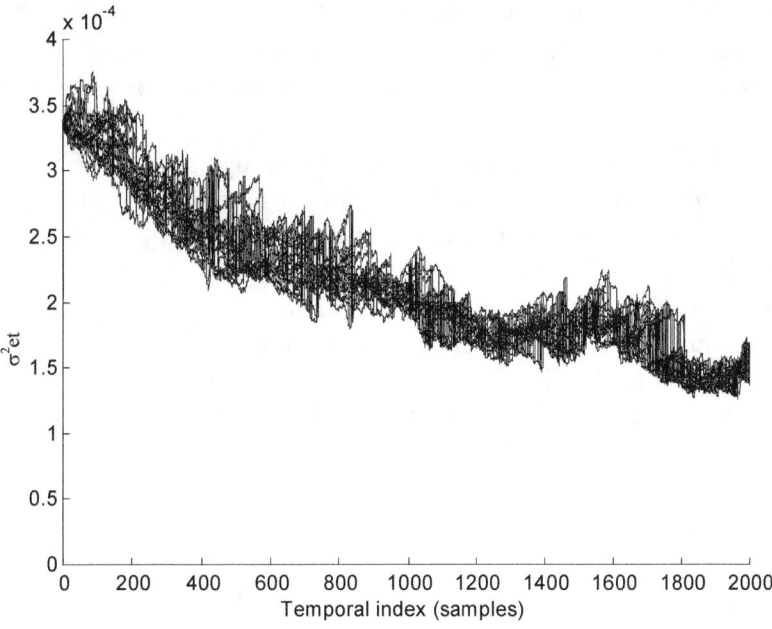

Fig. 8: *Temporal evolution of the variance σ^2_{et} obtained with the parameter $\alpha=1$, due to the only re-selection process. It is possible to observe how the action of the re-selection process prompts a choosing of the samples associated to a gradually decreasing value of the variance, eliminating the trajectories which are too far away from the optimal value.*

For the parameter $\sigma^2_{\phi_e}$ (variance of the *random walk* applied to the noise variance ϕ_{et}) we noticed that a too small value does not allow us to obtain a sufficient differentiation in the values of ϕ_{et}. The consequence consists in the fact that the selection phenomenon of the samples generated by using more appropriate parameters is damaged, leading to less valid results. An acceptable quantity order can be $\sigma^2_{\phi_e} \sim 10\text{-}5$.

A good functioning of the algorithm hinges on the choice of the initial value of the input noise variance to the model σ^2_{et}. If a too small value is chosen, the N samples, generated in the filtering phase, are very close to the starting noisy signal and therefore the restoration obtained is not very different from the input signal. On the contrary, if the initial value chosen is too high for σ^2_{et}, the automatic adjusting process towards smaller values is quite slow and does not manage to make a correction in due times. The resulting signal is therefore very different from the noisy signal available, and there is the risk of obtaining a restoration which does not resemble the original signal very much. Although it is not possible to establish a precise value to initialize the variance σ^2_{et}, since it depends on the signal and on the order of the model used, it was observed that a good starting value can be $\sigma^2_{et} = 10\text{-}3$.

By exploiting the option of selecting the samples associated with the best parameters and therefore of the automatic setting up of the filter it is possible to create a *tracking* procedure for the choice of the initial values to assign to the model coefficients.

A rather significant estimation may be obtained by using the above-described filtering algorithm which was obtained with the Monte Carlo method with the Rao-Blackwell strategy and considering the values generated for the coefficients contained in the state vector at a certain time t.

3.   Validation

In this section some experiments, carried out to test the validity of the algorithms presented in the previous chapters and to evaluate the performance level obtained by applying the implementations of the various models, will be illustrated. The algorithms were tested on different types of musical genres, on artificially corrupted signals and real recordings.

3.1   Numeric indexes used

The serious problem concerning the employment of real material is represented by the lack of an original to refer to for the qualitative evaluations, since the listening impressions can vary in a scarcely controllable way. Perceptive experiments showed that the given evaluation changes depending on whether the subject is willing to give more attention to the musical signal or to the noise.

Since the restoration algorithms work more efficiently in the case of synthetic degradation, it is highly preferable to have a non-degraded original, on which to base comparisons with the restored signal regarding both perceptive value and from a strictly numerical point of view. In this section, some deviation indexes (SNR relations and spectral distances) will be used to evaluate the measures with objectivity.

Although acknowledging that nothing can substitute direct listening to evaluate the performances obtained, we nevertheless decided to use numeric indexes to verify whether a correlation between these indexes and perceptive impressions existed, thus allowing for an objective comparison between the different methods used.

In the time domain a simple check, which can be carried out on a clean signal, consists in calculating the relation of the maximum deviation between the restored signal and the clean signal and the maximum value of the deviation in the noisy signal (which can be considered equal to $3\sigma_e$, in the case of Gaussian noise with variance σ_e^2) expressed in decibels. However it is convenient to use the average value of this index calculated on segments of 20ms.

We therefore define:

$$MxD_{seg} = \frac{1}{M} \sum_{i=1}^{M} 20 \log_{10} \left(\frac{\max|d_i|}{3\sigma_i} \right) \qquad (106)$$

in which d_i indicates the difference between the restored signal and the original and σ_i indicates the standard deviation of the noise, calculated for the i-nth segment, while M represents the number of segments used.

Some useful information can also be obtained by calculating the input signal-to-noise ratio:

$$SNR = 10 \cdot log_{10} \left(\frac{\sigma_{orig}^2}{\sigma_d^2} \right) \qquad (107)$$

where σ_{orig}^2 and σ_d^2 are the powers of the original signal and of the residual noise (difference between the original and restored). Even for this index a segmental definition is useful:

$$SNR_{seg} = \frac{1}{M} \sum_{i=1}^{M} SNR_i \qquad (108)$$

in which SNR_i represents the signal-to-noise ratio calculated for the *i*-nth signal segment of 20ms.

Ultimately an index connected to the harmonic components of the quantities being considered was calculated:

$$SpD = \frac{10}{ln(10)} \cdot \int_0^{2\pi} \left(ln \left| \frac{S_{orig}}{S_{rest}} \right|^2 \right)^2 \frac{d\theta}{2\pi} \qquad (109)$$

In this case S_{orig} and S_{rest} represent the periodograms of the original signal and of the restored signal (calculated with the Welch method). Note that the equation (109) represents a spectral distance definition often employed in the field of vocal signal coding (Gray and Markel, 1976).

A segmental definition considering the average value calculated on signal segments of length equal to 20ms, is useful for this index as well, since the human ear is more sensitive to local values than to total values:

$$SpD_{seg} = \frac{1}{M} \sum_{i=1}^{M} SpD_i \qquad\qquad (110)$$

A future development in this sense could concern the experimentation of indexes which take into account psychoacoustic aspects (for example the masking effect or the average response curves of the human ear). However, the scarce aptitude to really describe the sound quality shown by *classic instrumental* measures as those above-described need be emphasised.

The attempt of choosing the filtering modality or the parameters concerning the filter only according to numeric indexes is not always an optimal choice. Direct listening of the difference between the restored signal and the original signal can be very useful to evaluate the effectiveness of the restoration operation. In this way, it is possible to establish whether only the noisy component or even some musical component was eliminated.

3.2   Validation of the methods in the frequency domain

Given the abundance of the parameters available, only a representative subset was chosen. In particular, the noisy versions with a higher signal-to-noise ratio were tested with every implemented filter, leaving all the other parameters unchanged (using the default values). In this way, an attempt to optimize at best the working of each filter was made.

From the tests carried out it is evident how in the employment of the Wiener and pseudo-Wiener filter there is a significant quantity of residual noise modulated around the signal frequencies. A strong presence of musical noise can be observed. The EMSR filter has a decidedly minor musical noise (perceptively non-relevant for high SNR), but it is necessary to regulate the parameter α in order to find a right compromise between the noise attenuation and the transient distortion. The filter EMSR <*alpha*> shows a higher noise attenuation without causing audible distortions.

The filter based on the psychoacoustic model works properly with signals with low SNR, which are therefore very disturbed, while for the signals with high SNR the restoration obtained is sometimes particularly pervasive and eliminates many components, above all at high frequencies of the original signal, as can be observed listening to the difference between the noisy input signal and the restored output signal.

From the tests carried out important indications can be obtained: two fundamental parameters of the dimension and of the analysis of the windows overlapping were changed. As already discussed in Canazza *et al.* (1999), a high overlapping, exceeding 80%, has no advantages for the noise reduction. On the contrary the advantage consisting in window dimension increase is evident, provided that also the overlapping increases (to make the forwarding step constant in terms of samples number). However a bigger window has the disadvantage of leaving a grater quantity of residual noise. The study concluded that the dimensions of 4096 and 8192 samples represent a good compromise (a sampling frequency of 44.1 kHz was considered).

In the presence of a high noise quantity in comparison with the signal (SNR < 15dB) the need of overestimating the noise print to carry out an effective reduction arises. In particular, the musical noise (present also with the EMSR <*alpha*> filter) can be *coloured* by acting on the high frequencies overestimation. Modifying the last Barks of the noise print it is possible to make this defect less audible.

In highly degraded pieces, better results can sometimes be obtained by operating two subsequent filtering processes. In fact, when using a single filtering, it is necessary to resort to an overestimation of the noise print in order to avoid that the signal obtained still preserves a perceptively appreciable quantity of noise. This overestimation implies however an excessive filtering, which also modifies the useful signal, condition which should be avoided, as far as possible, in a good restoration. Instead, operating with two subsequent phases it is possible to obtain a restoration able to preserve the useful signal. The data listed in Table 1 concern a single use of the filters.

Another adjustable parameter concerning the psychoacoustic model used consists in the number of bands into which the spectrum is divided (25, 50 or 100 bands). During the listening, it is possible to observe that the implemented psychoacoustic model works much better with a division which employs only a low number of bands, since when using a high number of bands there is a significant amount of residual noise, which can be attenuated only by increasing the overestimation of the noise print. This operation implies an evident degradation of the original signal. Therefore the division of the spectrum into only 25 bands for each restoration was preferred. The last choice consists in using Kokkinakis's or Beerends psychoacoustic model. In the results obtained using these two models no differences were found.

Table 1 shows the list of numeric indexes discussed above and associated with the restorations obtained by using the different filters in the frequency domain, optimized from time to time in the choice of values to attribute to each parameter. The musical signal consists in a fragment (6 sec) of a song for singing voice only (*Tom's Diner* by Suzanne Vega, from the CD A&M, 5136-2), which had been previously corrupted by adding white Gaussian noise with different intensity and therefore with different SNR levels (respectively: 10 dB, 20 dB e 30 dB) to the signal. The choice of this piece was made in order to compare these results with those obtained from various prestigious international laboratories (the Sound Processing Group of Cambridge University, the Sound Engineering of Danzica University), which often use this piece for their experiments.

	MxDseg	SNR	SNRseg	SpD	SpDseg
Noisy 10db	/	10.00 dB	4.69 dB	575.33 dB	381.28 dB
Wiener	-5.23 dB	16.29 dB	11.16 dB	433.00 dB	259.94 dB
Pseudo-Wiener	-1.96 dB	13.04 dB	7.81 dB	509.05 dB	324.11 dB
EMSR	-5.69 dB	46.91 dB	11.74 dB	411.95 dB	249.24 dB
EMSR alpha	-6.52 dB	17.69 dB	12.63 dB	387.44 dB	229.03 dB
Psycho-acoustic	-6.77 dB	16.49 dB	12.84 dB	88.21 dB	38.39 dB

	MxDseg	SNR	SNRseg	SpD	SpDseg
Noisy 20db	/	20.00 dB	14.61 dB	396.6 dB	238.37 dB
Wiener	-4.13 dB	25.15 dB	19.98 dB	280.83 dB	148.45 dB
Pseudo-Wiener	-1.6 dB	22.63 dB	17.36 dB	342.34 dB	195.39 dB
EMSR	-4.08 dB	25.33 dB	20.17 dB	264.00 dB	140.62 dB
EMSR alpha	-4.48 dB	25.71 dB	20.67 dB	244.50 dB	126.27 dB
Psycho-acoustic	-3.04 dB	23.54 dB	19.47 dB	90.24 dB	36.39 dB

	MxDseg	SNR	SNRseg	SpD	SpDseg
Noisy 30db	/	30.00 dB	24.54 dB	251.82 dB	132.56 dB
Wiener	-1.86 dB	32.81 dB	27.84 dB	193.13 dB	92.31 dB
Pseudo-Wiener	-1.09 dB	32.04 dB	27.01 dB	209.17 dB	102.78 dB
EMSR	-3.11 dB	34.10 dB	29.30 dB	149.35 dB	66.59 dB
EMSR alpha	-3.39 dB	34.33 dB	29.61 dB	134.88 dB	57.78 dB
Psycho-acoustic	-1.04 dB	32.29 dB	27.77 dB	65.32 dB	21.45 dB

Tab. 1: *Value of the different quality indexes in function of the restorations obtained using the different filters in the frequency domain*

Figure 9 shows the gain trend introduced by each filter at the varying of the noisy signal SNR in the restoration of the above mentioned piece. The term gain indicates the difference between the restored signal SNR and the input signal SNR. However, the SNR value is merely indicative of the restoration quality: for example , in the psychoacoustic filter which exploits noise masking on the part of the useful signal, the residual noisy – but masked – component contributes to decrease this index even if it is not perceptible. For all input SNR, the EMSR *<alpha>* filter can be said to have performed well.

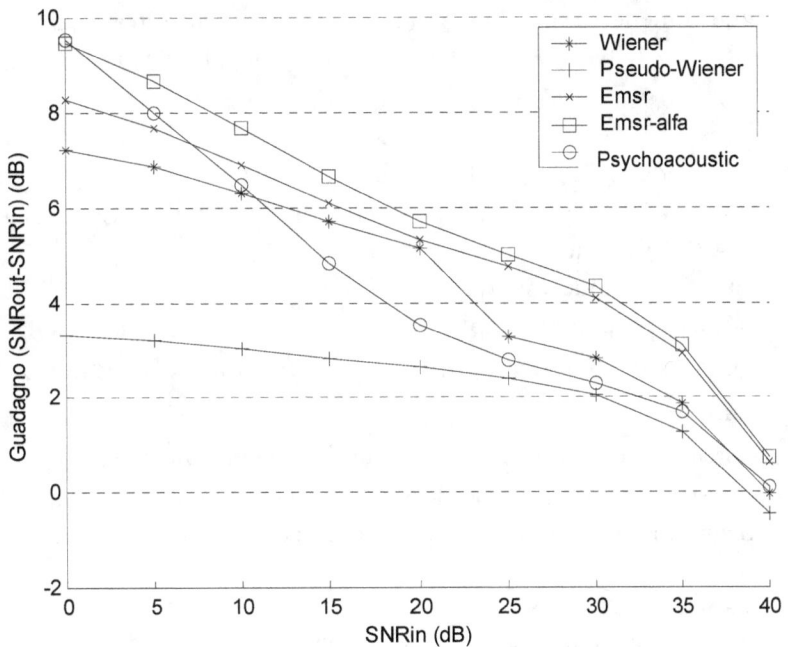

Fig. 9: *Gain trend introduced by the filters in the frequency domain at the varying of the input SNR. The best gain of the Emsr <alpha> filter can be observed for all the SNR$_{in}$.*

3.3 Validation of the methods in the time domain

Among the algorithms previously presented, the Extended Kalman Filter is the one that was also considered for the reduction of local disturbances (*clicks*). The testing conditions were therefore chosen in a way to be as realistic as possible, with the simultaneous presence of background noise and impulsive disturbances. A fragment sampled directly from a damaged vinyl record was adopted as a pattern for the clicks. The duration of the single click is of 16 samples; Figure 10 exemplifies its form. The original signal, retrieved from the CD, is a piano piece present in the *Prelude n. 15* by Fryderyk Chopin (performed by Maurizio Pollini, Deutsche Grammophone, 1975). The duration is of about 12 seconds. White (Gaussian) noise was added to it in order to obtain a SNR equal to 35dB.

Furthermore the piece was corrupted with a click train having variable amplitude (between 8% and 25% of the full scale signal and randomly distributed (with mutual distances included between 0.2 and 1.5 seconds).

With a single passage only the most evident clicks were removed, by using the arithmetic mean of two *forward* and *backward* filtering processes, obtained separately by using the EKF version (with the EWLS) setting $\gamma = 1\text{-}10^{-6}$ and $\lambda = 0.97$. The functioning parameters are:

- AR model order: $p = 14$ ($q_1 = 2$)
- thresholds of the impulses detector: $\mu = 5$ (for the Forward) and $\mu = 4$ (for the Backward)
- clustering windows: $cw_1 = cw_2 = 2$
- bootstrap of 3300 samples
- initial estimation of the variance $\hat{\sigma}_e^2(0)$: Youle-Walker out of 3300 samples
- initial estimation of the background noise variance: $\hat{\sigma}_z^2(0) = 0$
- filtering processes carried out starting from the extremes (the beginning of the piece for the Forward and the end for the Backward)
- stability bond : ON
- dynamic overestimation of $\hat{\sigma}_z^2(0)$: OFF

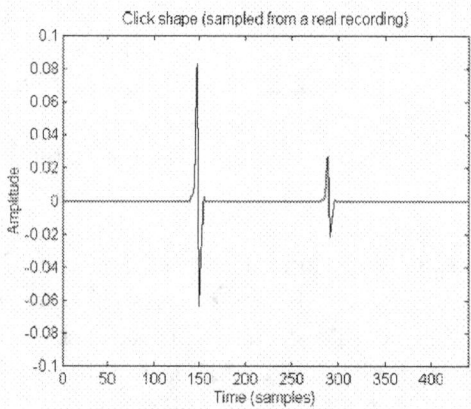

Fig. 10: *Detalis of the trend of the used clicks*

Figures 11 and 12 illustrates the comparison between the degraded signal and the trends of the restoration and of the original signal (indicated with dotted lines), in particular, a part in which the presence of a click is evident, was framed. The temporal window is of 0.1s. Note that this disturbance is to be considered a particularly difficult case, since it is characterised by an amplitude comparable to that (local) of the analysed signal, and also in virtue of its positioning in an area of high gradient. As we can observe, the click removal was carried out by the EKF with a significant accuracy, proven by the fact that the original signal track is well overlapped to that of the restoration.

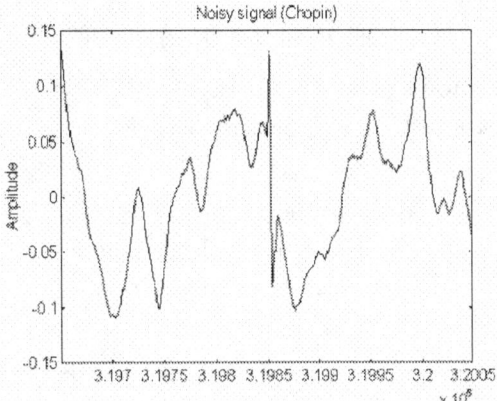

Fig. 11: *Detail of a click addition (Chopin).*

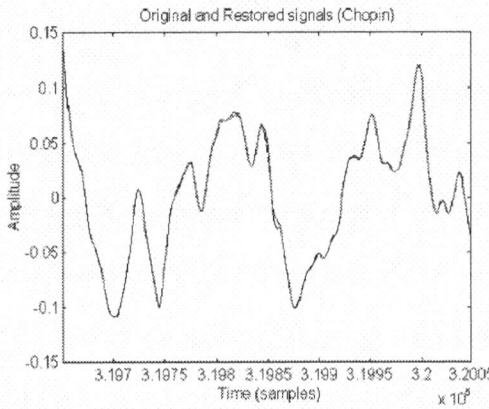

Fig. 12: *Detail of a click removal (Chopin).the original piece and the restored piece are overlapped, to prove the effectivness of the algorithm (EKF) employed.*

In order to compare the effectiveness of the two algorithms operating in the time domain, presented in section 2, and to compare their performance with those obtained with the filters in the frequency domain (*see Tab. 1*), the numeric indexes associated with the restorations of the *Tom's Diner* piece by Suzanne Vega, previously corrupted by adding white Gaussian noise with different SNR levels (10 dB, 20 dB e 30 dB) to the original signal, are listed in Table 2. The parameters were formulated as follows:

- for the Extended Kalman Filter: $p = 8$, $\mu = 0$, $cw_1 = cw_2 = 4$, bootstrap of 2000 samples, $\hat{\sigma}_e^2(0)$: Youle-Walker out of 2000 samples, $\hat{\sigma}_z^2(0)$=real value (known), stability bond: ON, dynamic overestimation of $\hat{\sigma}^2{}_z(0)$: OFF

- for the Monte Carlo algorithm with Rao-Blackwell strategy: $p=8$, $\sigma^2{}_{mc}=10^{-4}$, $\hat{\sigma}_e^2(0) = 10^{-3}$, $\sigma^2{}_{\phi_c} = 10^{-5}$, $\alpha = 0.9998$, $\hat{\sigma}_z^2(0) = $ real value (known), bootstrap of 2000 samples, number of samples for each block $T = 2000$, number of particles generated $N = 70$, number of realisations of the trajectory estimation $M = 5$.

	MxDseg	SNR	SNRseg	SpD	SpDseg
Noisy 10db	/	10.00 dB	4.69 dB	575.33 dB	381.28 dB
EKF	-2.17 dB	13.25 dB	7.96 dB	494.25 dB	313.99 dB
Montecarlo Rao-Blackwell	-5.91 dB	15.19 dB	9.82 dB	227.13 dB	152.81 dB

	MxDseg	SNR	SNRseg	SpD	SpDseg
Noisy 20db	/	20.00 dB	14.61 dB	396.6 dB	238.37 dB
EKF	-6.39 dB	26.70 dB	21.85 dB	118.46 dB	29.26 dB
Montecarlo Rao-Blackwell	-4.78 dB	24.01 dB	19.22 dB	120.15 dB	49.21 dB

	MxDseg	SNR	SNRseg	SpD	SpDseg
Noisy 30db	/	30.00 dB	24.54 dB	251.82 dB	132.56 dB
EKF	-3.89 dB	34.57 dB	29.60 dB	92.00 dB	21.36 dB
Montecarlo Rao-Blackwell	-1.87 dB	31.18 dB	26.41 dB	105.03 dB	37.82 dB

Tab. 2: *Value of the different quality indexes in function of the restorations obtained by using filters in the time domain*

The graphic of Figure 13 (compare this with Figure 9) shows the gain trends in terms of difference between output SNR and input SNR, introduced by the Extended Kalman Filter and by the algorithm with the Monte Carlo method and Rao-Blackwell strategy, in the restoration of the piece by Suzanne Vega. Note the scarce effectiveness of the EKF for input SNR values inferior to 15dB.

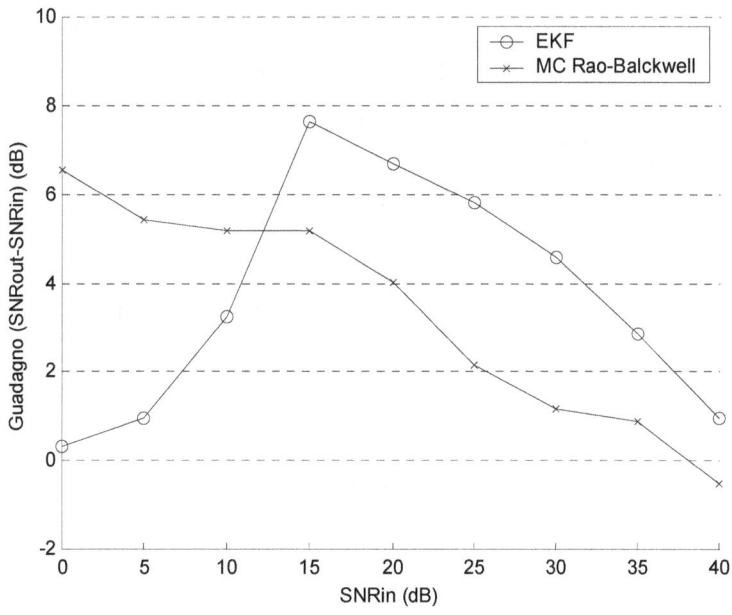

Fig. 13: *Gain trend introduced in the time domain by the filters at the varying of the SNR in input*

4. An example of restoration by means of analysis by synthesis

An approach (particularly fit in the electro-acoustic music field) which differs from the methods in the frequency and time domains, foresees the study of the compositional model and of its analogue production system aiming at separating the *useful signal* from the *noise*, intended in the double meaning of non-musical signal (broadband or impulsive noise) and the deviation from the model proposed by the composer. In this way the two noise typologies can be studied both for restorative purposes (the more information we have about the noise, the more precise the parameters of the models used in the restoration can be) and for historical-musicological purposes (the noise can reveal information on the equipment used by the laboratory where the piece was produced).

In the electro-acoustic music field some compositions are made up of elementary audio signal combinations (sinusoids) with proper amplitude and frequency modulated through time-varying indexes. If the mathematic model used by the composer and the technology adopted in the production of the piece are known, we can hypothesise to reconstruct the musical signal by synthesizing the different frequency components according to the "analysis by synthesis" method (Gabrielsson, 1997). This foresees the following steps:

a) measuring of the audio object parameters: duration, spectrum, intensity and envelope. The variables to study depend on the work hypothesis and on the technical possibilities of the measurement instruments;

b) selection and analysis of the most important variables in order to characterize the work, in function of the mathematical model adopted by the composer;

c) numerical interpretation of the data: choice of the different representations and temporal scales;

d) musical work synthesis according to the data emerging from the measures;

e) study of the relation between the original work and the synthesized work, in order to focus on the correlations between different parameters, to single out the most important variables and determine the variables to eliminate;

f) repetition of the procedure (steps b÷e) till the result converges.

In order to illustrate this methodology, it was applied to *Analogique B* by Iannis Xenakis. In *Formalized Music* (Xenakis, 1992), Iannis Xenakis describes the technical procedures and the mathematical structure of the work, from which it is however impossible to perform the re-recording of the work. Therefore the analysis by synthesis strategy was applied, by modelling the composition directly on the analysis of the sound material.

The composition, for magnetic tape, carried out in 1959 at the GMR studies ("Groupe de Recherches Musicales") of Paris, is based on a granular synthesis model, inspired by the Gabor and Moles theories. For exemplification, an abstract (6 s) taken from a pattern supplied by Ricordi-Salabert was examined.

In this way Xenakis described the granular synthesis model for Analogique B:

> «all sound, even all continuous sonic variation, is conceived as an assemblage of a large number of elementary grains adequately disposed in time. [...] In the attack, body, and decline of a complex sound, thousands of pure sounds appear in a more or less short interval of time, Δt»
> (Xenakis, 1992, p. 43).

Xenakis organized the sound grains in *screens*, a sort of temporal intervals inside which hundreds of grains are distributed in a statistical way. It came out that the composer exclusively used sinusoids, generated through the use of analogue oscillators. The last of the grains, according to the author's intention – "to grasp the depth of the present" – can be ascribed to 40 ms, but there are no certain elements to exclude different values. As far as intensity is concerned, that is, the amplitude of the grains, Xenakis wrote that he would use only four levels (Xenakis, 1992, p. 106).

A sound grain is endowed with its own frequency, amplitude, duration and envelope. The minimum duration necessary to distinguish the pitch depends on the signal frequency: a threshold of 13 ms for high frequencies, which increases up to 45 ms for the low frequencies, is necessary. Under these limits the grain is perceived as a *click*.

The work was divided into the following phases: determination of the frequencies, the times and the amplitudes of each grain, calculation of the filter parameters for the singling out of the envelopes, re-synthesis of the piece by using the results of the analysis, in order to highlight some aspects of the noise ascribable to the technology of the time through subtraction with the historical recording, and therefore to model the behaviour of the historical equipment.

For the frequencies discrimination, autocorrelation was used in order to discriminate the sinusoid frequencies constituting the grains. According to an analysis carried out on the whole signal it was observed that beyond 10 kHz there are no grains. Therefore only the frequencies up to this value were considered. Furthermore, sounds with an amplitude inferior to -96 dB were considered as noise.

To reconstruct the envelopes two different waterfalls filter typologies were used: low-band, centred on each value which emerged from the analysis in order to obtain grains concerning the respective frequencies; low-pass to obtain the envelope of the grains.

Limiting the selectivity of the filters to obtain a settling time compatible with the grain duration was necessary, in order to reach a correct evaluation of the signal amplitude. FIR filters with Kaiser window, (more linear in the response in band in comparison with elliptic IIR filters) parameterized as in Table 3, were used.

	Pass band filters	Pass low filters
Centre band frequency:	frequencies emerged from the analysis;	
Pass band frequency	$F_{p1} = F_{cb} - 1$ Hz; $F_{p2} = F_{cb} + 1$ Hz;	$F_p = 107,7$ Hz;
Cut frequency at −6 dB:	$F_{c1} = F_{cb} - 23$ Hz; $F_{c2} = F_{cb} + 23$ Hz;	$F_c = 125$ Hz;
Band width:	BW = 46 Hz;	
Obscure band frequencies:	$F_{s1} = F_{cb} - 46$ Hz; $F_{s2} = F_{cb} + 46$ Hz;	$F_s = 177,6$ Hz;
Attenuation in obscure band:	$R_s \geq 40$ dB;	$R_s \geq 20$ dB;
Beta:	$\beta = 3.395$;	$\beta = 3.395$;
Filter order:	2188;	657;
Ripple in band:	$R_p < 1$ dB;	$R_p < 3.551$ dB;

Tab. 3: *Parameters used in the filter planning*

To correctly evaluate the signal envelope, two different modalities on the basis of the signal frequencies were carried out: interpolation of the maximum peaks and low-pass filtering. The threshold frequency between the two modalities was set at 140 Hz. Under this frequency, the filtering was not able to eliminate the grain audio frequency and therefore it did not correctly supply the envelope.

The analysis allowed us to obtain the starting and ending samples of the grains and their amplitude.

The estimated envelopes are delayed – about 7.25 ms (320 samples) – in comparison with the real ones and the signal is proportionally scaled in amplitude. Some corrections were therefore carried out:

1) the envelopes reconstructed through interpolation of the peaks, produced only for frequencies inferior to 140 Hz, were delayed of 320 samples so that they preserve the right relative temporisation in comparison with those obtained through filtering;

2) the envelopes obtained through filtering were amplified by a value equal to the relation between the signal maximum before the low-pass filtering and the envelope maximum, thus restoring the real amplitude values.

Therefore the maximum peaks were calculated for each envelop.

The following step consisted in calculating the duration of each grain: we chose – trying different hypothesis on purposely synthesised grains- to determine the starting and endings times, evaluating the times at the hemi value.

The calculation of the samples which represent the starting and ending times of a grain depended on the mutual position of the relative maximum peaks, as schematized in Figure 14.

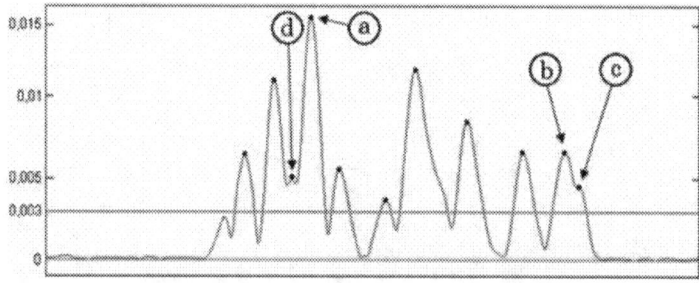

Fig. 14: *Different peak typology (underlined in black) of the grains wraps.*

For each peak we searched for the first sample to its left and the first to its right whose amplitudes were close to half of that of the considered peak. If on one hand for the *a* peaks the values can be easily found moving to and fro the vector containing the samples, on the other hand as far as the peaks of the *b* typology are concerned this method allows only for the singling out of the starting sample. Ultimately, as far as the grains of type *b* are concerned both the values cannot be determined. Therefore the peak value was introduced instead of the missing samples and the grains in the matrix containing the data deriving from the analysis were associated with an index associated to the typologies of Figure 14.

For each frequency the grains *a* were considered and those whose duration were inferior to 10 ms – minimum value necessary to perceive the pitch were eliminated. As for the remaining ones, their average duration was calculated: with this value the duration of the grains of type *b*, *c* and *d* was therefore approximated.

6 different amplitude levels were defined, with an interval of 6 dB between them. These values were associated to the musical notations *ff*, *f*, *mf*, *mp*, *p* and *pp*. since the maximum and minimum values of the signal resulted -20 dB and -50 dB respectively, the levels -20 dB, -26 dB, -32 dB, -38 dB, -44 dB and -50 dB were chosen. The grain amplitudes were therefore approximated to a closer level.

The synthesis of the piece was carried out by using the frequencies remaining after the analysis (66 arranged in the interval between 47 and 6331 Hz) with the duration partially corrected, so that all the grains were constituted of an integer of sinusoid semi-periods. This expedient sensibly decreased the introduction of perceivable spikes. The variation introduced was, in the worse case of 2.5 ms for grains at low frequency.

For each grain a sinusoid of the desired frequency and duration was modelled. Trapezoidal envelopes were applied to these grains, with peaks of 1 ms – therefore similar to the theoretical rectangular trend, foreseen by Xenakis.

Figure 15 shows the sonograms of the original (a) and of the synthesized signal (b). The synthesized signal was delayed by a dozen of milliseconds in comparison with the first, however without altering the mutual position of the grains.

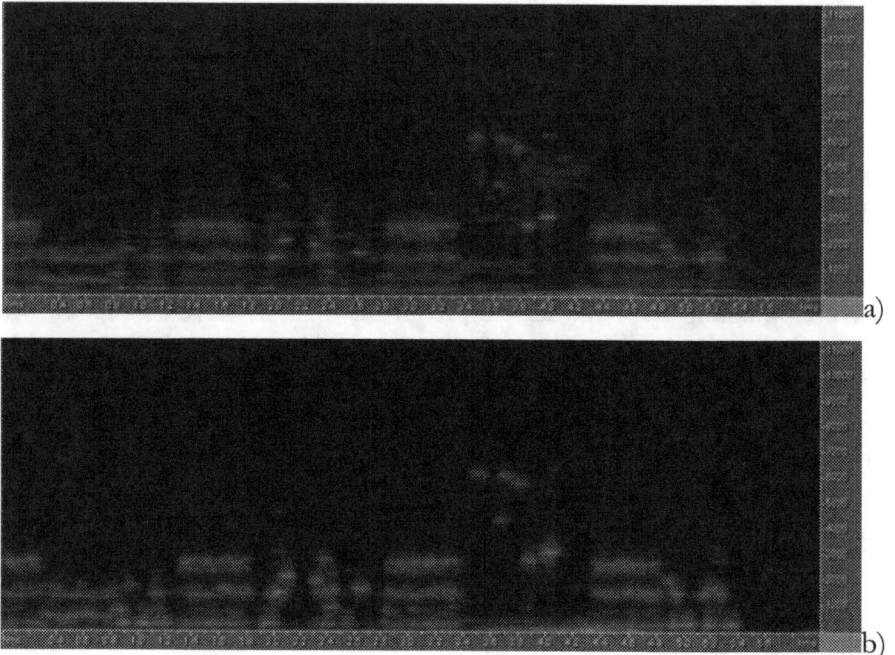

Fig. 15: *Sonograms of the original (a) and of the synthesized signal (b) —Hanning windows, length of 1024 samples , 60% overlapping.*

The useful signals (the grains, in the case examined) of the two images are very similar, but in Figure 15b we can note the absence of noise. The signal obtained through the subtraction between the original and the synthesized signal contains the noise to be ascribed to: a) the imperfections and/or to the aging of the carrier; b) the insufficiencies of the system used by the composer. Hypothesizing to have access to the musical instruments (electronics, in the case examined) used by the composer, they can be modelled (waveguide synthesis) and the models can be used for the synthesis of the piece. In this way the synthesized sound coincides exactly (in the ideal case) to the original sound, as it was produced by the composer.

In order to check the information *external* to the signal about the work, some statistics were carried out *(fig. 16)*. The first *(fig. 16)* concerns the distribution of the grains of type *a* only (see. *fig. 14*) in comparison with the durations estimated, the mean value is 30.4 ms. This distribution is thought to belie the hypothesis that the author used grains of the same length, equal to 40 ms. The second statistic (*fig. 16b*) concerns the distribution in dB of all the grains in comparison with the amplitudes, estimated during the analysis. We can clearly infer that not only 4 levels of intensity are present, hence the choice, during the signal synthesis, to consider a higher number of levels.

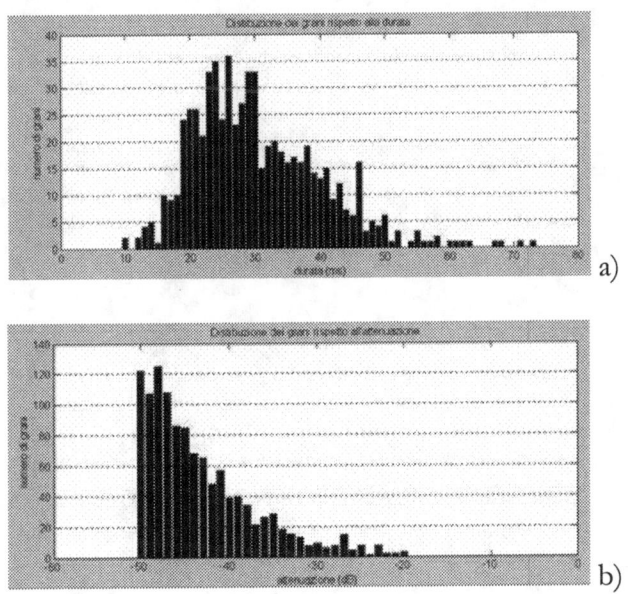

a)

b)

Fig. 16: *a) Distribution of the grains in comparison with the estimated duration; b) Distribution of the grains in comparison with the estimated amplitude.*

In the examined segment, the maximum grain density (maximum number of grains for each second) calculated in the signal interval with the highest concentration (between 4.31 and 4.85 seconds) is equal to 370, coherent with one of the values indicated by Xenakis (1992, p. 105).

Conclusions

The problem of audio restoration has been treated proposing algorithms, which are based on different approaches. The algorithms in the frequency domain, representing an innovation in comparison with the traditional methods, have been illustrated. Besides these algorithms, psychoacoustic considerations have been added thus enabling for the improvement of the obtained results. Ultimately, methods in the time domain have been presented, which basing themselves on signal models make it possible to obtain a significant improvement of the audio signal quality.

In order to compare the performances obtained by different restoration algorithms, we have decided to introduce some numeric indexes. However, we have to stress the fact that it is not possible to obtain a real numeric measure able to summarize the restoration result completely. In this sense nothing can replace direct listening.

Among the different methodologies proposed, the filters operating in the time domain, at the expenses of a careful formulation of the different parameters and of a high computational complexity, allow us to obtain a restoration in which the degradation (impulsive disturbances and global noise) is attenuated without invalidating the useful component of the audio document with a low pass effect which, on the contrary, is usually present in the algorithms operating in the frequency domain. The scarce effectiveness of the EKF, in comparison with the Monte Carlo method with Rao-Blackwell strategy for values of input SNR inferior to 15dB is to be stressed: the behaviour is the opposite for values higher than SNR_{in}.

The filters in the time domain make it possible to operate in real time, with a reduced number of parameters to regulate. In this domain the EMSR <*alpha*> allow us to obtain a high SNR without the introduction of musical noise. In presence of a SNR minor to 15dB, the need of overestimating the noise print and to carry out an effective reduction arises. In the highly degraded pieces appropriate combinations of waterfalls filters can also be studied. In this sense, we can consider carrying out an processing with an EMSR <*alpha*> (introducing an attenuation to the noise mask) and a second one with the psychoacoustic filter. In this way the risk of removing the useful components of the signal is reduced.

However we cannot define an *optimal* restoration strategy since every operator follows his own subjective aesthetical principles and is conditioned by the aesthetics of the historical time in which he is working.

Bibliography

Anderson, B. D. O., and Moore, J. B. (1979). *Optimal filtering*. Englewood Cliffsm, NJ: Prentice-Hall.

Bari, A., Canazza, S., De Poli, G., and Mian, G. A. (2001). Toward a methodology for the restoration of electro-acoustic music. *Journal of new music research*, *30(4)*, 365-374.

Beerends, J. G., and Stemerdink, J. A. (1992). A perceptual audio quality measure based on psychoacoustic sound representation. *Journal of Audio Engineering Society*, *40(12)*, 963-978.

BPN029. (2000). *EBU Report on the Subjective Listening Tests of Some Commercial Internet Audio Codecs. Contribution from EBU Project Group B/AIM.*

Boll, S. F. (1979). Suppression of acoustic noise in speech using spectral subtraction. *IEEE transactions on acoustics, speech, and signal processing*, *27(2)*, 113-120.

Boll, S. F. (1991). Speech enhancement in the 1980s: noise suppression with pattern machine. In S. Furui and M. Sondhi (eds.), *Advances in speech signal processing* (pp. 309-326). New York: Marcel Dekker.

Canazza, S., De Poli, G., Maesano, S., and Mian, G. A, (1999). On the performance of a noise reduction technique based on a psychoacoustic model for the restoration of old audio recordings. In *Proceeding of Diderot Forum*, Vienna, 2-4 December 1999 (pp. 29-35).

Canazza, S., De Poli, G., Mian, G. A., and Scarpa, A. (2001). Objective comparison of audio restoration methods based on short time spectral attenuation. In *Proceeding of "Science and technology for the safeguard of cultural heritage in the mediterranean basin"*, Alcalá de Henares, Spain, 9-14 July 2001 (pp. 173-174).

Canazza, S., De Poli, G., Mian, G. A., and Scarpa A. (2002). Comparison of different audio restoration methods based on frequency and time domains with applications on electronic music repertoire. In *Proceeding of International computer music conference*, Goteborg, Sweden, 16-21 September 2002 (pp. 104-109).

Cappé, O. (1991). *Noise reduction techniques for the restoration of musical recordings* (text in French), Ph. D. thesis. France: École Nationale Supérieure des Télécomunications, Paris.

Cappé, O. (1994). Elimination of the musical noise phenomenon with the Ephraim and Malah noise suppressor. *IEEE transactions on speech, audio processing, 2(2)*, 345-349.

Carrey, M. J., and Buckner, I. (1976). A system for reducing impulsive noise on gramophone reproduction equipment. *The radio electronic engineer, 50(7)*, 331-336.

Casella, G., and Robert, C. P. (1996). Rao-Blackwellization of sampling schemes. *Biometrika, 83*, 81-94.

Czyzewski, A. (1997). Learning algorithms for audio signal enhancement. Part I: Neural networks implementation for the removal of impulsive distortions. *Journal of Audio Engineering Society, 45(10)*, 815-831.

Doucet, A., de Freitas, N., Murphy, K., and Russel, S. (2000). Rao-Blackwellised particle filtering for dynamic Bayesian networks. In A. Doucet, N. de Freitas, and N. Gordon (eds.), *Proceeding of Sixteenth conference on uncertainty in artificial intelligence*, Stanford, USA, 2000. San Francisco, CA: Morgan Kaufmann.

Doucet, A., Godsill, S. J., and West, M. (2000). Monte Carlo filtering and smoothing with application to time-varying spectral estimation. In *Proceeding of IEEE International conference on acoustics, speech, and signal processing*, Istanbul, Turkey, June 2000 (vol. 2, pp. 701-704).

Efron, A. J., and Jeen, H. (1992). Pre-whitening for detection in correlated plus impulsive noise. In *Proceedings ICASSP*, San Francisco, USA, 1992 (p.te II, pp. 469-472).

Ephraim, Y., and Malah, D. (1983). Speech enhancement using optimal non-linear spectral amplitude estimation. In *Proceeding of IEEE of*

International conference on acoustics, speech, and signal processing, Boston, USA, 1983 (pp. 1118-1121).

Ephraim, Y., and Malah, D. (1984). Speech enhancement using a minimum mean-square error short-time spectral amplitude estimator. *IEEE transactions on acoustics, speech, and signal processing, 32(6)*, 1109-1121.

Ephraim, Y., and Malah, D. (1985). Speech Enhancement using a minimum mean-square error log-spectral amplitude estimator. *IEEE transactions on acoustics, speech, and signal processing, 33(2)*, 443-445.

Esquef, P. A. A., Välimäki, V., and Karjalainen, M. (2002). Restoration and enhancement of solo guitar recordings based on sound source modeling. *Journal of the Audio Engineering Society, 50(4)*, 227-236.

Etter, W. (1996). Restoration of a discrete-time signal segment by interpolation based on the left-sided and right-sided autoregressive parameters. *IEEE transactions on acoustics, speech, and signal processing, 44(5)*, 1124-1135.

Fong, W., Godsill, S. J., Doucet, A., and West, M. (2002). Monte Carlo smoothing with application to audio signal enhancement. *IEEE transactions on signal processing, 50(2)*, 438-448.

Fortescue, T. R., Kershenbaum, L. S., and Ydstie, B. E. (1981). Implementation of self-tuning regulators with variable forgetting factors. *Automatica, 17*, 831-835.

Friedlander, B. (1982). Lattice filters for adaptive processing. *Proceedings IEEE, 70(8)*, 829-867.

Gabrielsson, A. (1997). Music performance. In D. Deutsch (ed.), *The psychology of music*. New York: Academic Press.

Godsill, S. J. (1993). *The restoration of degraded audio signals*. Ph. D. thesis. England: Cambridge University Eng., Department Cambridge.

Godsill, S. J. (1998). *Bayesian enhancement of speech and audio signal in the presence of both impulsive and background noise.* Internal Report. England: Signal Processing and Communication Group, University of Cambridge.

Godsill, S. J., and Rayner, P. J. W. (1993). Frequency-domain interpolation of sampled signals. In *Proceeding ICASSP*, Minneapolis, USA, 1993 (vol. 1, pp. 209-212).

Godsill, S. J., and Rayner, P. J. W. (1995). A Bayesian approach to the restoration of degraded audio signals. *IEEE transactions on speech, and audio processing, 3(4)*, 267-278.

Godsill, S. J., and Rayner, P. J. W. (1998). *Digital audio restoration – A statistical model-based approach.* London: Springer-Verlag.

Grancharov, V., Samuelsson, J., and Kleijn, B. (2004). Noise-dependent postfiltering. In *Proc. IEEE Int. Conf. Acoustics, Speech, Signal Processing*, USA, 2004 (vol.1, pp. 457-460).

Grancharov, V., Samuelsson, J., and Kleijn, B. (2005). Improved Kalman Filtering for Speech Enhancement. In *Proc. IEEE Int. Conf. Acoustics, Speech, Signal Processing*, USA, 2005 (vol. 1, pp. 1109-1112).

Grancharov, V., Samuelsson, J., and Kleijn, B. (2006). On casual Algorithms for speech enhancement. *IEEE transaction on audio, speech, and language processing*, 14(3), 273-276.

Gray, A. H. Jr., and Markel, J. D. (1976). Distance measures for speech processing. *IEEE transactions on acoustics, speech, and signal processing, 28(4)*, 380-390.

Harvey, A. C. (1989). *Forecasting, structural time series models and the Kalman Filter.* Cambridge, UK: Cambridge University Press.

Huber, P. J. (1981). *Robust statistics.* New York: Wiley.

Jassem, A. J. E. M., Veldhuis, R. N. J., and Vries, L. B. (1986). Adaptive interpolation of discrete-time signals that can be modelled as autoregressive processes. *IEEE transactions on acoustics, speech, and signal processing, 34,* 317-330.

Kasparis, T., and Lane, J. (1993). Adaptive scratch noise filtering. *IEEE transactions on consumer electronics, 39(4),* 917-922.

Kinzie, G. R. Jr., Gravereaux, D. W. (1973). Automatic detection of impulse noise. *Journal of Audio Engineering Society, 21(3),* 331-336.

Lewis, F. L. (1986). *Optimal estimation.* New York: Wiley.

Lim, J. S., and Oppenheim, A. V. (1978). All-pole modelling of degraded speech. *IEEE transactions on acoustics, speech and signal processing, 26(3),* 197-210.

Lim, J. S., and Oppenheim, A. V. (1979). Enhancement and bandwidth compression of noisy speech. *Proceedings IEEE, 67(12),* 1586-1604.

Ma, N., Bouchard, M., and Goubran, R. A. (2006). Speech enhancement using a masking threshold constrained Kalman Filter and its heuristic implementations. *IEEE transaction on audio, speech, and language processing,* 14(1), 19-32.

McAulay, R. J., and Malpass, M. L. (1980). Speech enhancement using a soft-decision noise uppression filter. *IEEE transactions on acoustics, speech and signal processing, 28(2),* 137-145.

McAulay, R. J., and Quartieri, T. F. (1986). Speech analysis/synthesis based on a sinusoidal representation. *IEEE transactions on acoustics, speech and signal processing, 34(4),* 744-754.

Montresor, S., Valiere, J. C., Allard, J. F., and Baudry, M. (1990). The restoration of old recordings by means of digital techniques. In *Actes du 88ème AES,* 13-16 march, Montreux.

Montresor, S., Valiere, J. C., Allard, J. F., and Baudry, M. (1991). Evaluation of two interpolation methods applied to old recordings restoration. In *Actes du 89ème AES,* Lyon.

Niedzwiecki, M. (1989). Steady-state and parameter tracking properties of self-tuning minimum variance regulators. *Automatica, 25(4)*, 597-602.

Niedzwiecki, M. (1997). Identification of time-varying processes in the presence of measurement noise and outliers. In *Proceedings of IFAC workshop on new trends in design of control systems*, Smolenice, Slovak Republic (vol. 4, pp. 1765-1770).

Niedzwiecki, M. (2000). *Identification of time-varying processes.* New York: Wiley.

Niedzwiecki, M., and Cisowksi, K. (1996). Adaptive scheme for elimination of broadband noise and impulsive disturbance from AR and ARMA signals. *IEEE transactions on signal processing, 44(3)*, 528-537.

Nieminen, A., Miettinen, M., Heinonen, P., and Nuevo, Y. (1987). Music restoration using median type filters with adaptive filter substructures. In V. Cappellini and A. Constantinides (eds.), *Digital signal processing – 87*; Amsterdam, North-Holland.

Orcalli, A. (2006). Orientamenti ai documenti sonori. In S. Canazza and M. Casadei Turroni Monti (eds.), *Ri-mediazione dei documenti sonori.* Udine: Forum.

Ptas, I., and Venetsanopoulos, A. N. (1990). *Nonlinear digital filters.* Kluwer Academic Publishers.

Tsoukalas, D., Mourjopoulos, J., and Kokkinakis, G. (1997). Perceptual filters for audio signal enhancement. *Journal of Audio Engineering Society, 45(1-2)*, 22-36.

Turner, S. (1997). *Demonstrating Harmony: Some of the Many Devices Used To Produce Lissajous Curves Before the Oscilloscope.* New York: Rittenhouse.

Valiere, J. C. (1991). *La restauration des enregistrements anciens par traitement numériques – Contribution á l'étude de quelques techniques recents.* Ph. D. thesis. France: Université du Maine, Le Mans.

Van Trees, H. (1968). *Decision, estimation and modulation theory* (part 1). New York: Wiley.

Vaseghi, S. V. (1988). *Algorithms for restoration of archived gramophone recordings.* Ph. D. thesis. England: University of Cambridge.

Vaseghi, S. V., and Rayner, P. J. W. (1988). A new application of adaptive filters for restoration of archived gramophone recordings. *Proceedings ICASSP*, New York, USA, 11-14 April 1988 (pp. 2548-2551).

Vaseghi, S. V., and Rayner, P. J. W. (1990). Detection and suppression of impulsive noise in speech communication systems. *Proceedings IEEE*, *137(1)*, 38-46.

Veldhuis, R. (1990). *Restoration of lost sample in digital signals.* New York: Prentice Hall.

Vermaak, J., Andrieu, C., Doucet, A., and Godsill, S. J. (1999). *Non-stationary Bayesian modelling and enhancement of speech signals.* Tech. Rep. CUED/F-INFENG/TR.351 (Dept. Eng., Cambridge University, England).

Wiener, N. (1949). *Extrapolation, interpolation, and smoothing of stationary time series with engineering applications.* Cambridge, MA: MIT Press.

Wolfe, P. J., and Godsill, S. J. (2000). The application of psychoacoustic to the restoration of musical recordings. In *Proceedings of the 108th convention of the Audio Engineering Society*, Paris, 19-22 february 2000 (pp. 254-273).

Xenakis, I. (1992). *Formalized music.* Stuyvesant, NY: Pendragon Press.

Ydstie, B. E., and Sargent R. W. H. (1986). Convergence and stability properties of an adaptive regulator with variable forgetting factor. *Automatica, 22,* 749-751.

Index